1st International Conferences on Physics Issues (ICoPIs 2021)

Journal of Physics: Conference Series Volume 2126

Palu, Indonesia
28 August 2021

ISBN: 978-1-7138-5097-7
ISSN: 1742-6588

Printed from e-media with permission by:

Curran Associates, Inc.
57 Morehouse Lane
Red Hook, NY 12571

proceedings.com

Some format issues inherent in the e-media version may also appear in this print version.

Institute of Physics
Dirac House, Temple Back
Bristol BS1 6BE UK

Phone: 44 1 17 929 7481
Fax: 44 1 17 920 0979

techtracking@iop.org

Additional copies of this publication are available from:

Curran Associates, Inc.
57 Morehouse Lane
Red Hook, NY 12571 USA
Phone: 845-758-0400
Fax: 845-758-2633
Email: curran@proceedings.com
Web: www.proceedings.com

TABLE OF CONTENTS

Journal of Physics: Conference Series **2126** (2021) 011001 doi:10.1088/1742-6596/2126/1/011001

PREFACE
The 1st International Conference on Physics Issues (ICoPIs2021)
Universitas Tadulako, Palu, Indonesia
28 August 2021

The International Conference of Physics Issues (ICoPIs) is an international conference held by the Department of Physics Education, Faculty of Teacher Training and Education, Tadulako University. The 1st International Conference on Physics Issues (ICoPIs2021) was a virtual conference organized by the Department of Physics Education in the Faculty of Teacher Training and Eucation, Tadulako University on 28th August 2021. This conference covered a variety of works, including the works on 1) learning strategy, model, method, media and learning evaluation in physics education; 2) nanoscience and nanomaterials; 3) research result of physics; 4) strategy, model, method, media and evaluation in natural science education; and 5) general education. The conference offered extraordinary opportunities for educators, teachers, researchers, administrators and decision-makers to exchange and gain new insights on the ongoing research works and an opportunity to collaborate with like-minded researchers from across the world.

Theme of the conference was "Bounce Back Better: Enhancing Innovative Thinking Through Independent Campus-Freedom to Learn (MBKM) during Pandemic Era". This event took the theme of "bounce back better" as an attempt to be resilient during the current pandemic situation. This conference was a forum for discussing and gaining knowledge and ideas from educational practitioners and experts. Accordingly, we hope that through this information exchanges, we will be more prepared for unprecedented situations in the future.

On behalf of the committee, we would like to thank the Chancellor of Universitas Tadulako, Dean of the Teacher Training and Education Faculty, Head of the Mathematics and Natural Sciences Department, advisory board, steering committee, all committee members, reviewers, presenters, and participants. Recognition should go to the ICoPIs2021 Team members who have all worked extremely hard for the details of important aspects of the conference programs and social activities. Special thanks are also given to the Proceedings of the IOP Conference. We, on behalf of the ICoPIs 2021 committee, would like to thank all parties for their participation in supporting this publication. We hope to see you all at the next conference.

Kind regards,

Dr. Nurjannah, M.Pd.

Chair of the ICoPIs 2021

ORGANIZING COMMITTEE

Person Responsible : Dr. Sahrul Saehana, M.Si
Chair : Dr. Nurjannah, M.Pd
Secretary Delthawati Isti Ratnaningtyas, M.Pd.
Treasurer Wahyuni N. Laratu, M.Pd.
Muh. Zaky, M.Pd

Publication Division Members

1. Ni Made Wiwik Astuti, M.Ed.
2. Ketut Alit Adi Untara, M.Pd.
3. Misbah, S.Pd., M.Pd. (ULM)

IT Division Members

1. Nurgan, M.Pd.
2. Dian Fitrasari, M.Si
3. Muh. Sarif, M.Pd.
4. Syamsuriwal, M.Pd.
5. Gustina, M.Pd.

Event Division Members

1. Novia, S.Pd., M.Pd (Universitas Tadulako)
2. Miftah, M.Pd.
3. Muh. Jarnawi, M.Pd.
4. Ielda Paramitha, M.Pd.
5. Aqidatul Izza

Conference Support Members

1. I Wayan Roby Yanto Maleaki, S.Pd.
2. Agnes Jois Palamba', S.Pd.
3. Sutriani
4. Susanti.
5. Rini Oktavia
6. Akbar Birrkahfi
7. Dhimas Arif Nurfachri
8. Andriyaztika Lala

Reviewer Team Members :

1. Dr. Achmad Samsudin, M.Pd (Universitas Pendidikan Indonesia)
2. Dr. Eng. Osi Arutanti, M.Si (LIPI Serpong)
3. Dr. Bebeh Wahid Nuryadin, M.Si (UIN Sunan Gunung Djati)
4. Dr. Ade Yeti Nuryantini, M.Si (UIN Sunan Gunung Djati)
5. Dr. Ketang Wiyono, M.Pd (Universitas Sriwijaya)
6. Muhammad Nur Hudha, M.Pd (Universitas PGRI Kanjuruhan Malang)
7. Dr. Jusman Mansyur, M.Si (Universitas Tadulako)
8. Riser Fahdiran, M.Si (Universitas Negeri Jakarta)
9. Ni Made Wiwik Astuti, M.Ed (Universitas Tadulako)

Editor Team Members :

1. Dr. Jusman Mansyur, M.Si (Universitas Tadulako)
2. Dr. Darsikin, M.Si (Universitas Tadulako)
3. Dr. Supriyatman, M.Pd. (Universitas Tadulako)

4. Dr. Haeruddin, M.Si (Universitas Tadulako)
5. Ketut Alit Adi Untara, M.Pd.
6. I Wayan Roby Yanto Maleaki, S.Pd. (Universitas Tadulako)
7. Agnes Jois Palamba', S.Pd (Universitas Tadulako)
8. Sutriani, S.Pd (Universitas Tadulako)
9. Misbah, M.Pd (Universitas Lambung Mangkurat)
10. Rini Oktaviani

Language Editor Team Members

1. Dian Fitrasari, M.Si (Universitas Tadulako)
2. Ni Made Wiwik Astuti, M.Ed (Universitas Tadulako)
3. Aqidatul Izza, S.Pd (Universitas Tadulako)
4. Nurul Imansari, M.A (Universitas Sulawesi Barat)
5. Desrin Lebagi, S.Pd., M.Pd (Universitas Tadulako)

Journal of Physics: Conference Series **2126** (2021) 011001 doi:10.1088/1742-6596/2126/1/011001

PHOTOGRAPHS

The 1st International Conference on Physics Issues (ICoPIs2021)
Universitas Tadulako, Palu, Indonesia
28 August 2021

Opening of the ICoPIs 2021 by the Vice Dean of General Affairs and Finance Abdul Kamaruddin, S.Pd., M.Ed., Ph.D

 2126 (2021) 011001 doi:10.1088/1742-6596/2126/1/011001

Opening of the ICoPIs 2021 by the head of the Sub-Department of Physics Education Dr. Sahrul Saehana, M.Si

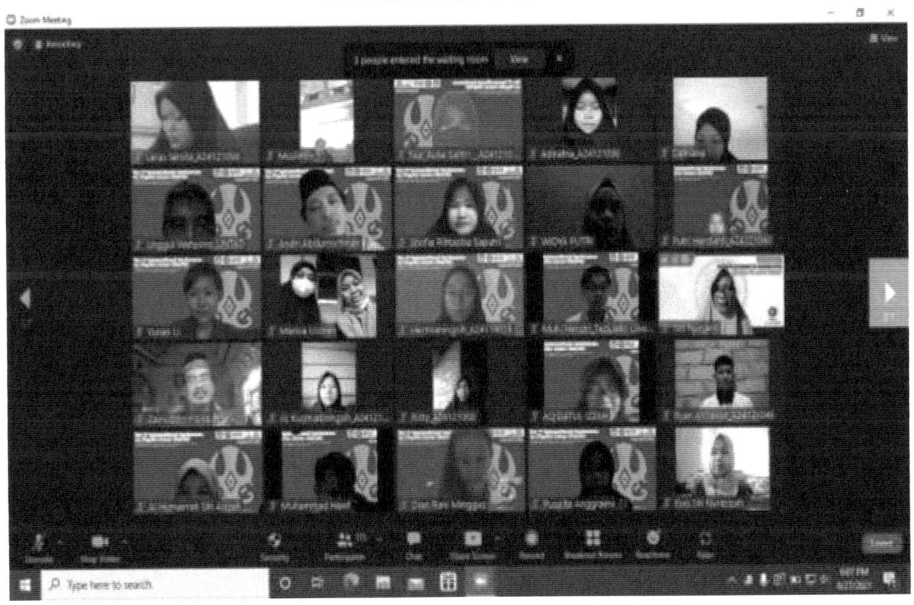

Photo Session

Journal of Physics: Conference Series **2126** (2021) 011001 doi:10.1088/1742-6596/2126/1/011001

Photo Session

Presentation by the keynote speaker Prof. Lilia Halim
(Universiti Kebangsaan Malaysia, Malaysia)

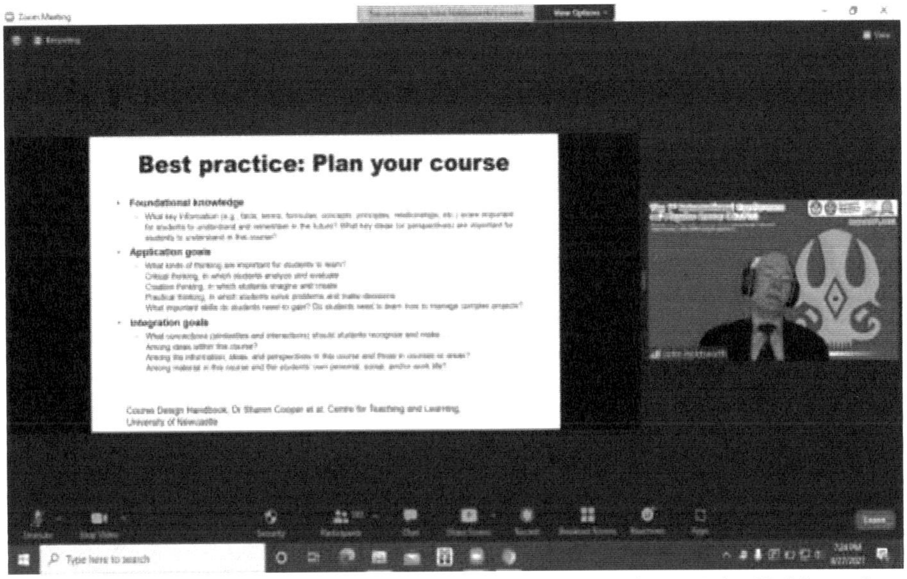

Presentation by the keynote speaker Associate Professor John Holdsworth
(The University of Newcastle, Australia)

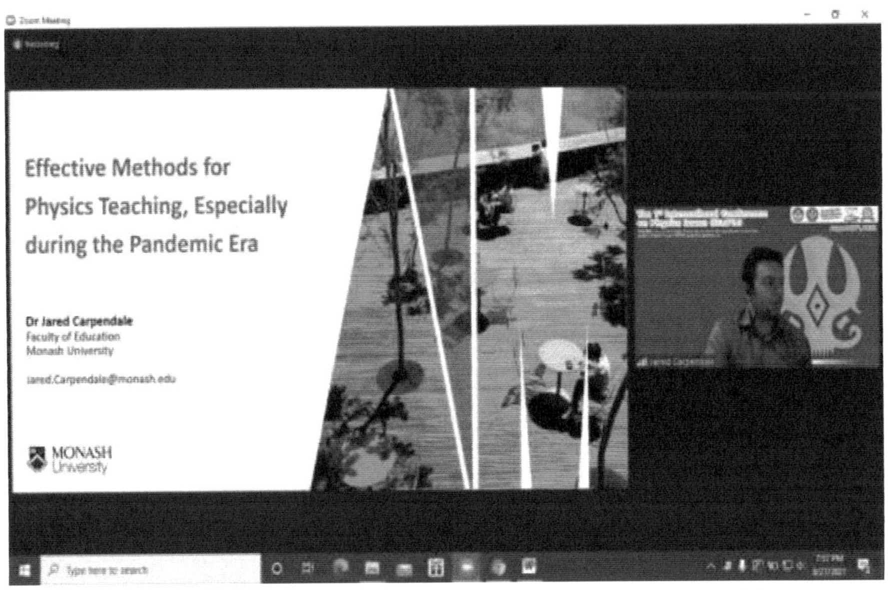

Presentation by the keynote speaker Dr. Jared Carpendale
(Monash University, Australia)

Journal of Physics: Conference Series **2126** (2021) 011001 doi:10.1088/1742-6596/2126/1/011001

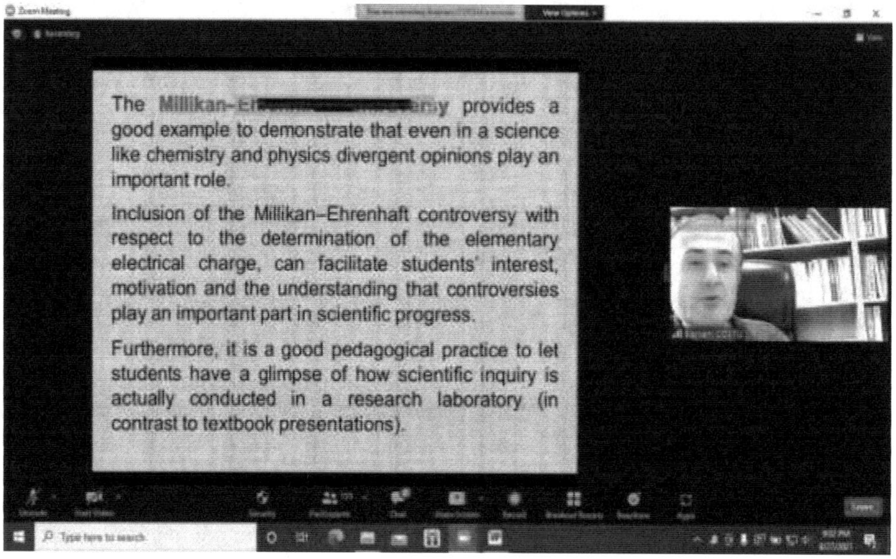

Presentation by the keynote speaker Prof. Bayram COŞTU
(Yıldız Technical University, Turkey)

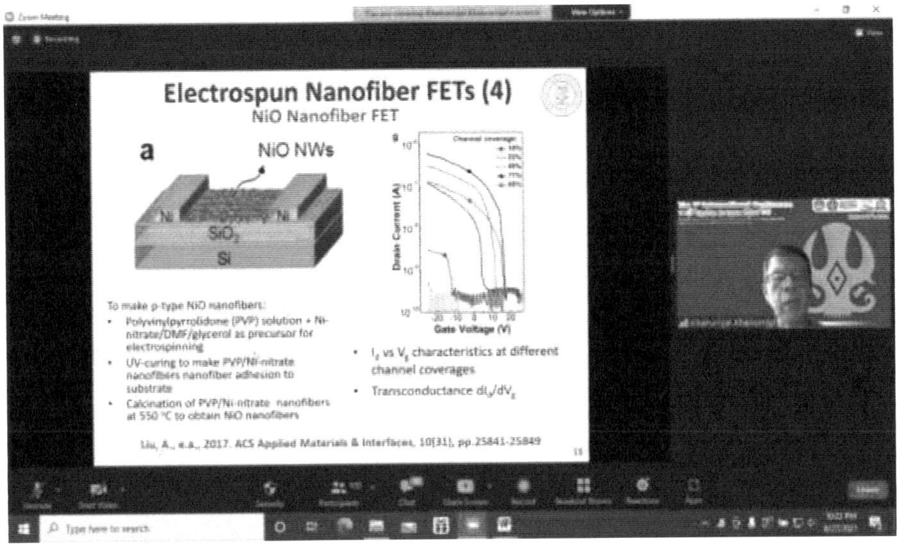

Presentation by the keynote speaker Prof. Dr. Eng. Khairurrijal (Bandung Institute of Technology, Indonesia)

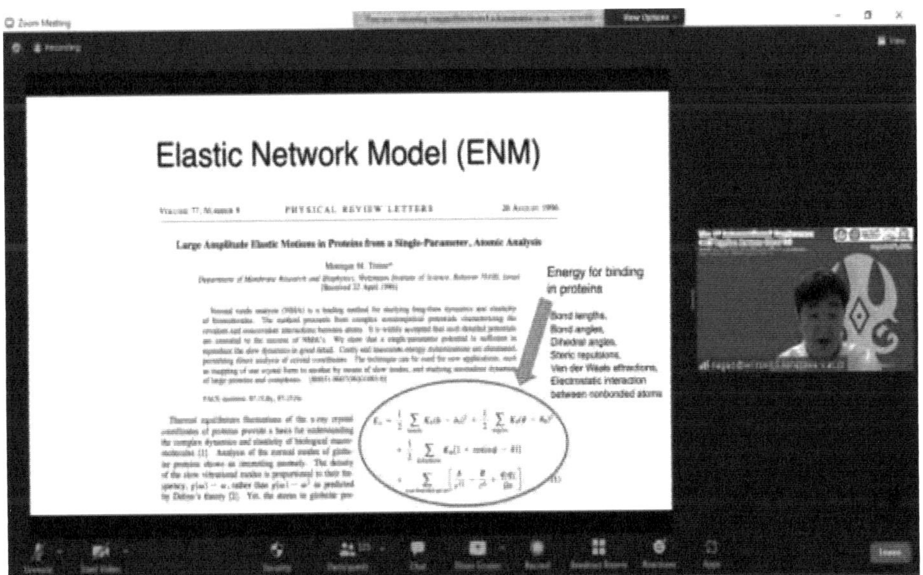

Presentation by the keynote speaker Prof. Hidemi Nagao (Kanazawa University, Japan)

Guest Speaker at the ICoPIs 2021 by Dr. Moh. Yusup, M.Pd (Sriwijaya University)

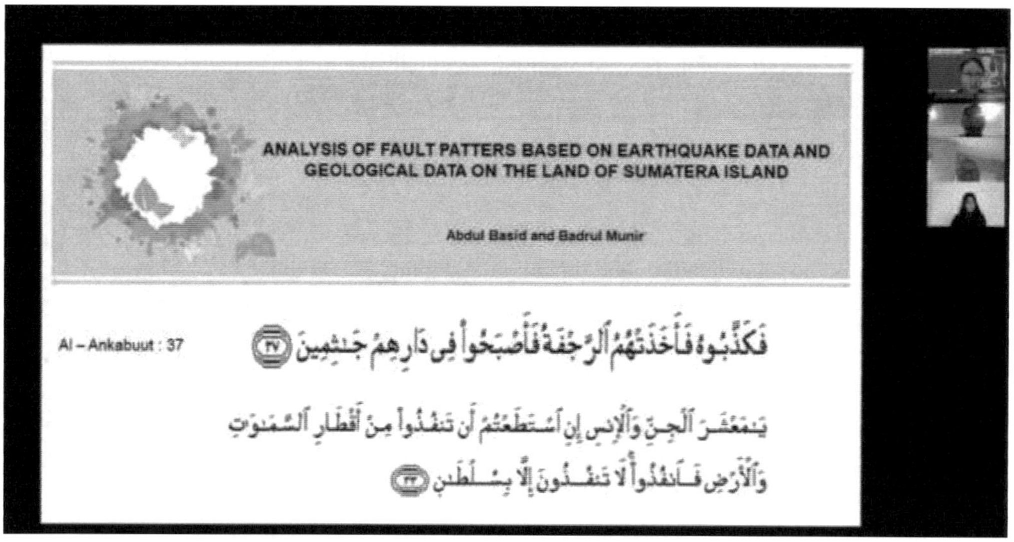

Parallel Session Presentation at the ICoPIs 2021

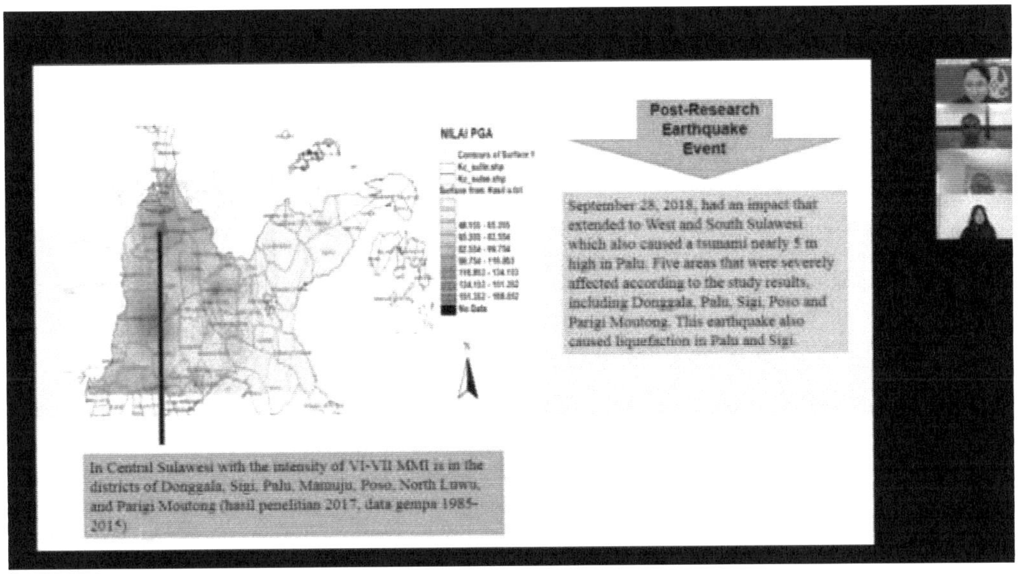

Parallel Session Presentation at the ICoPIs 2021

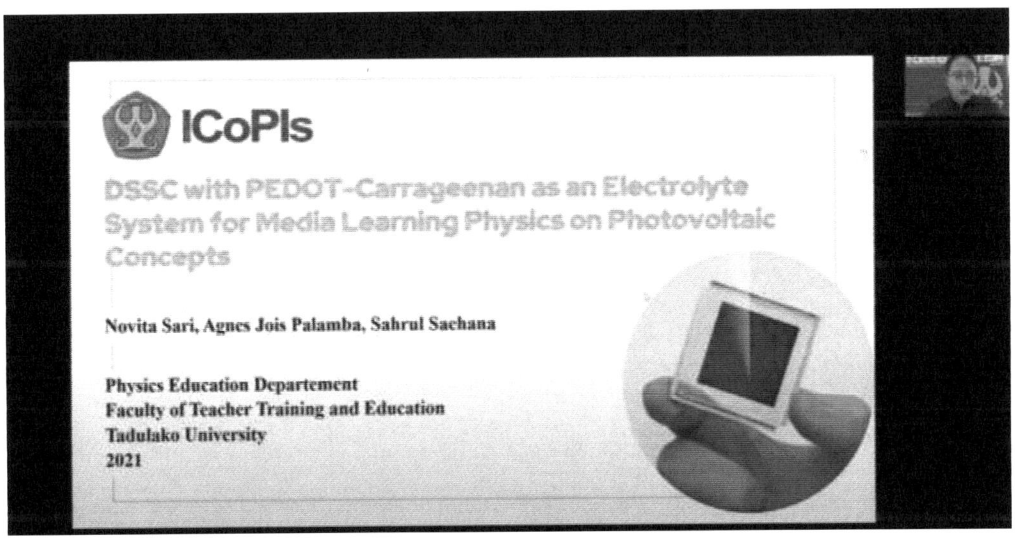

Parallel Session Presentation at the ICoPIs 2021

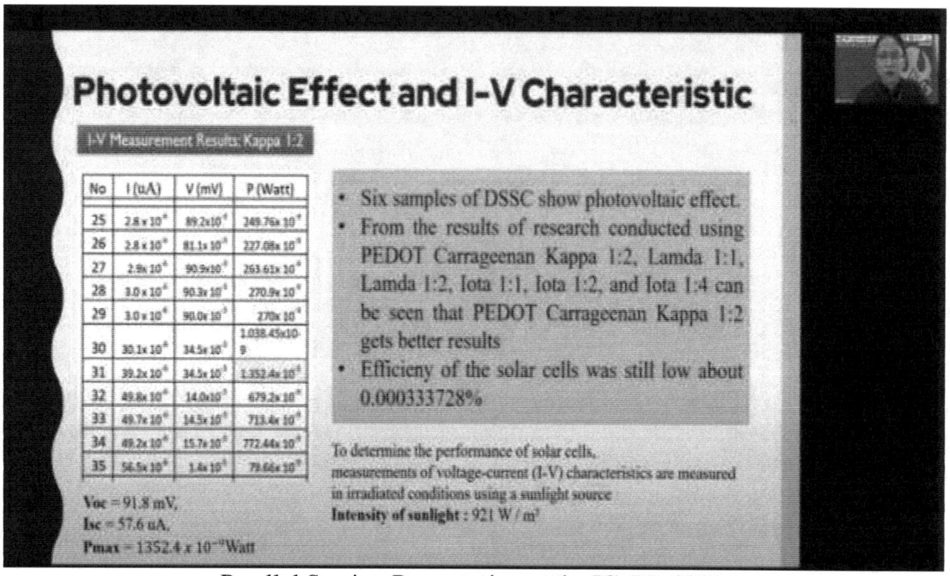

Parallel Session Presentation at the ICoPIs 2021

Parallel Session Presentation at the ICoPIs 2021

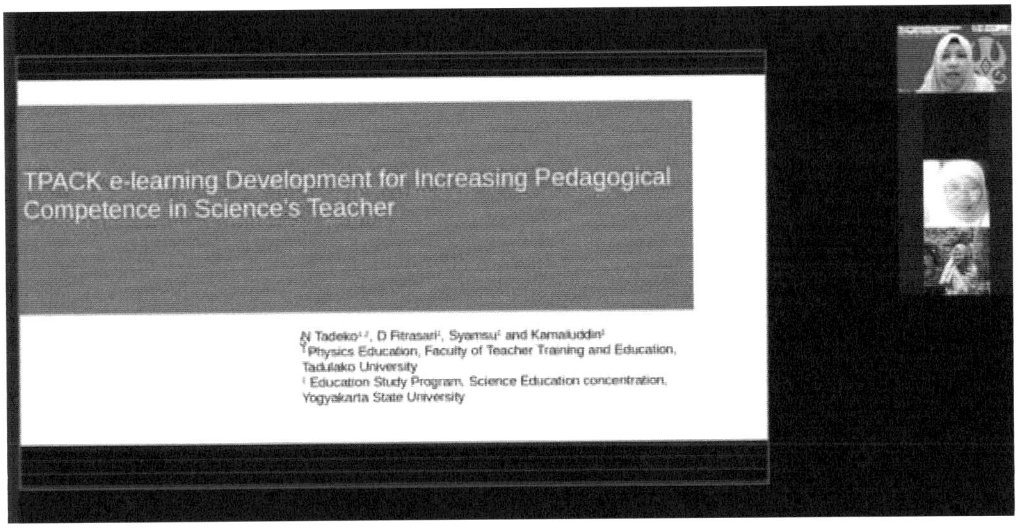

Parallel Session Presentation at the ICoPIs 2021

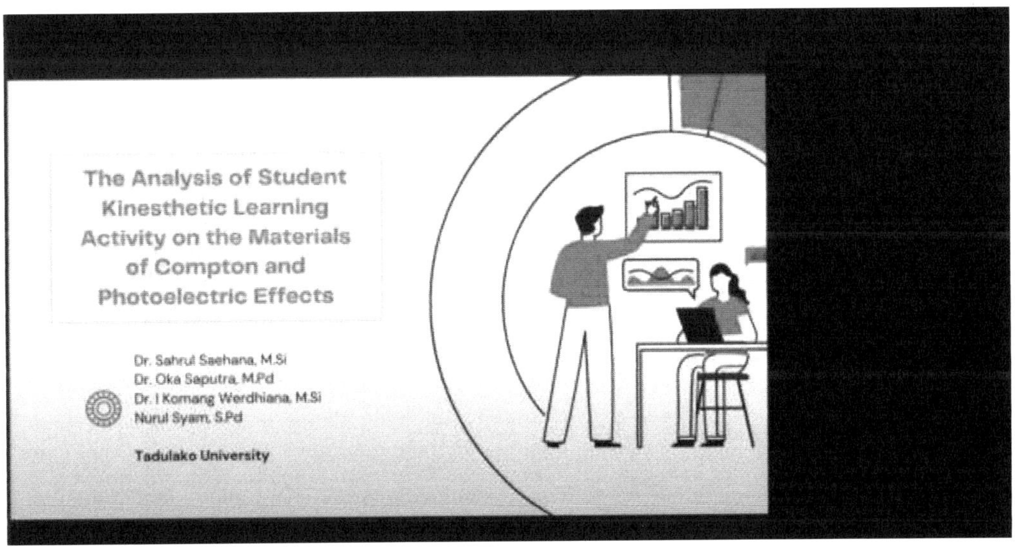

Parallel Session Presentation at the ICoPIs 2021

Parallel Session Presentation at the ICoPIs 2021

Parallel Session Presentation at the ICoPIs 2021

Parallel Session Presentation at the ICoPIs 2021

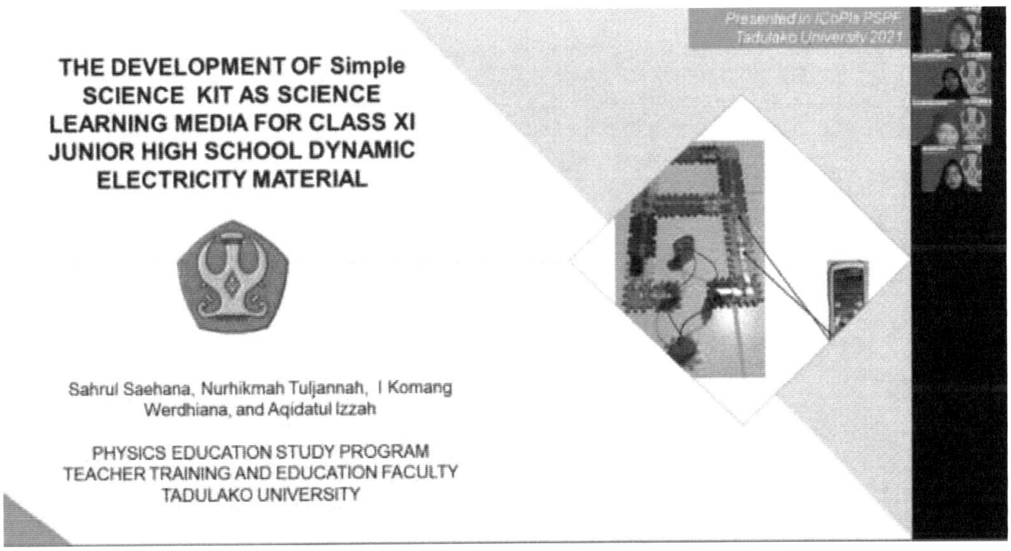

Parallel Session Presentation at the ICoPIs 2021

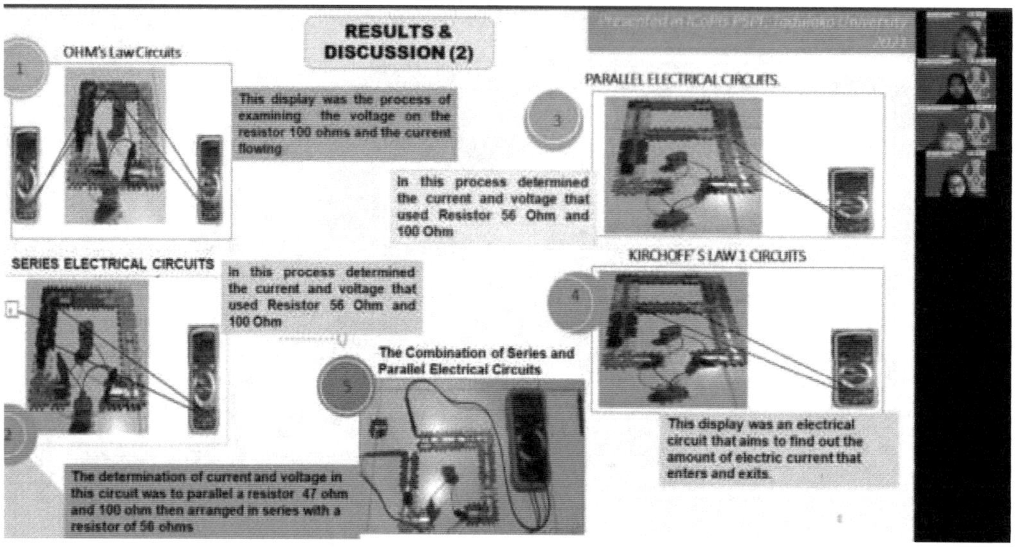

Parallel Session Presentation at the ICoPIs 2021

Parallel Session Presentation at the ICoPIs 2021

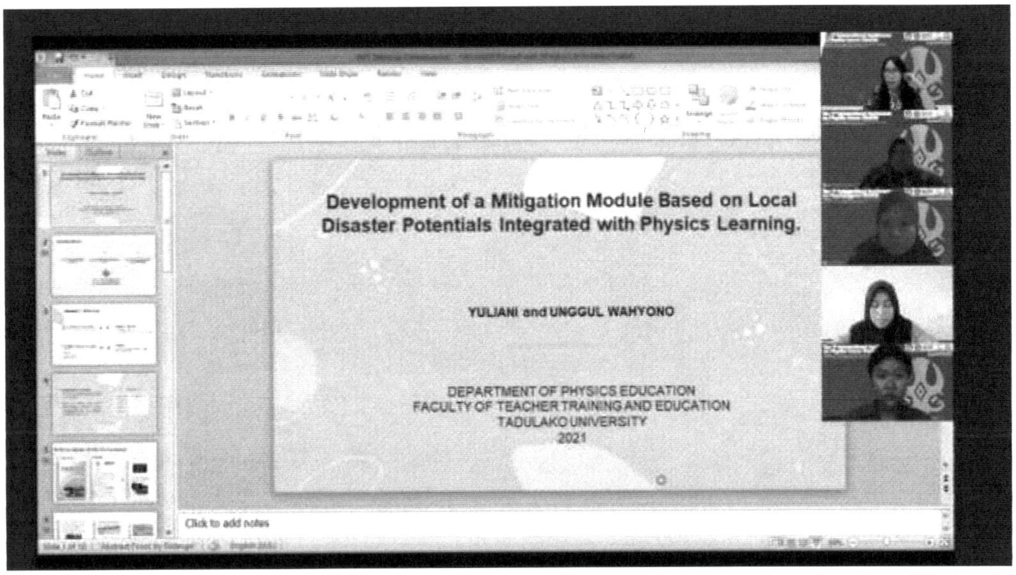

Parallel Session Presentation at the ICoPIs 2021

Parallel Session Presentation at the ICoPIs 2021

Parallel Session Presentation at the ICoPIs 2021

Parallel Session Presentation at the ICoPIs 2021

Peer review statement

All papers published in this volume of Journal of Physics: Conference Series have been peer reviewed through processes administered by the proceedings Editors. Reviews were conducted by expert referees to the professional and scientific standards expected of a proceedings journal published by IOP Publishing.

Journal of Physics: Conference Series **2126** (2021) 011002 doi:10.1088/1742-6596/2126/1/011002

Peer Review Declaration

All conference organizers/editors are required to declare details about their peer review. Therefore, please provide the following information:

- **Type of peer review:**
  The ICoPIs 2021 manuscripts were reviewed using double-blind peer review. One reviewer reviewed one manuscript. The number of papers submitted was 62 articles. The number of ICoPIs 2021 reviewers were 9 people. The reviewing processes were conducted via email or OCS. The manuscripts in ICoPIs 2021 have been through two stages of review. The first review stage involved three steps. First, manuscripts were distributed to the 10 reviewers. Second, the ICoPIs team checked the similarity of the manuscript. Third, the reviewed manuscripts consisting of the reviewer's detailed comments were returned to the authors, along with the review summary form and the similarity check results. The authors were requested to return the revision of their papers within a certain period of time. In the second review stage, the editorial team scrutinized the revised manuscripts with the summary review form and the similarity percentage. If the revision has followed the comments and suggestions from reviewers and the limitation of similarity, the manuscript would be forwarded for language and template check.

- **Conference submission management system:**
  OCS (https://fisika.fkip.untad.ac.id/icopis/)

- **The number of submissions received:**
  There were 62 articles submitted to the ICoPIs 2021

- **The number of submissions sent for review:**
  There were 56 articles reviewed by the reviewers of the ICoPIs 2021

- **The number of submissions accepted:**
  There were 37 articles accepted for recommendation/publication to JPCS IOP Publishing.

- **Acceptance Rate (Number of Submissions Accepted / Number of Submissions Received X 100):**
  59.68%

- **The average number of reviews per paper:**
  One article was reviewed 4 times; twice for content review and a similarity check, once for language and once more for the template.

- **The total number of reviewers involved:**
  There were 24 reviewers involved, consisting of 9 content reviewers, 10 editors (review content and templates), and 5 language reviewers.

- **Any additional info on the review process:**
  While the ICoPIs team waited for the manuscripts from the reviewers, we checked the similarity of the papers. We sent three documents to the authors for the first review stage, including the reviewed manuscript, the summary of the review form, and the result of a

similarity check. The authors sent the revised article along with the similarity check (maximum 20%). The manuscripts entered the second review stage. When a manuscript has followed the comments and suggestions from reviewers and was considered suitable for publication, the manuscript then proceeded to the editors for the template and language check. We did a similarity check once more during this phase but only for the manuscript with a high similarity percentage in the first review stage. So, one manuscript has been through a similarity check 2 to 3 times.

- **Contact person for queries:**
+628975586104 (Misbah)
Universitas Lambung Mangkurat
Misbah_pfis@ulm.ac.id

The calculation of uranium metallic Fuel (U-10%wtZr) cell with helium coolant using SRAC 2K6

Suci Claudia Putri[1], Menik Ariani[1], Idha Royani[1], Arsali[1], Fiber Monado[1*]
[1]Departement of Physics, Faculty of Mathematics and Natural Sciences, Sriwijaya University, Palembang, Indonesia, 30662

*Corresponding author: fibermonado@unsri.ac.id

Abstract. In this study, pin-shaped natural metallic uranium fuel cells with a diameter of 1.4 cm were designed with and without enrichment of 2-10% by using helium as a coolant. The calculations were conducted by using SRAC software and the results show that the greater the enrichment level, the more significant the value of the infinite-multiplication-factor (k_{inf}), meanwhile the conversion ratio is smaller. Basically, uranium exists in two isotopic forms namely, U-235 and U-238 which are transformed into other element such as Pu-239 when they both go through fission and transmutation reactions in the reactor core. Therefore, to obtain the optimum reactor core design, the performance information of nuclear fuel cells from the variations of burn-up time needs to be taken into consideration.

1. Introduction

Alternative sources of energy are needed to ensure sustainability in the future, and one of the promising alternatives is nuclear energy [1], which are produced from nuclear reactions. These reactions occur in a reactor core through merging (fusion) or fission, when two atomic particles collide to create a different product from the initial one [2]. Consequently, heat energy is produced in the core, hence a material with low neutron absorption, stable radiation, high temperatures, which is capable of absorbing the heat is required.

Also, nuclear fuel is the main component in a reactor obtained from natural isotopes such as uranium and thorium, which splits during a reaction with neutrons [3]. Meanwhile, the most widely used fuel is uranium, having two isotopes in nature, namely U-238 (99.3% by weight) and U-235 (0.7% by weight). The U-238 is a fertile material which does not split or undergo fission reactions with neutrons. However, U-235 is fissile, which means it is able to split and react directly with neutrons to undergo nuclear fission reactions [4]. Consequently, fertile material from natural uranium is used more effectively in fast reactors, while the neutrons produced react with the U-238 to create a new fissile material (Pu-239) [5].

A neutronic analysis includes the population of neutrons, flux, distribution, and its behavior in the reactor core. Inside the reactor core, there is a chain reaction of fission, and each reaction produces energy of 200 MeV. This reaction occurs when a neutron hits a heavy nucleus [6], however, it hardly occurs spontaneously. Therefore, comprehensive analyses such as neutronic, thermal-hydraulic, and safety are needed when designing a nuclear reactor. For example, the neutronic calculations are required to find out information such as multiplication factor values, neutron flux distribution, and power distribution [7]. Also, the characteristics of the neutrons need to be known because they are particles capable of causing fission reactions with fissile isotopes [8][9].

In this study, the fuel cells made from natural metallic uranium was calculated and the content consists of 99.3% uranium-238 and 0.7% uranium-235. The content of uranium-235 is increased through an enrichment process of 1-10% in order to utilize some nuclear reactors. Afterward, the uranium is combined with 10% zirconium (U-10%wtZr). Hence, this study compared survey parameters on fuel cell burn-up for 100, 120, 140, and 160 years. This comparison is carried out to determine the performance of nuclear fuel cells to ensure the continuous production of critical values at a predetermined burn-up time.

2. Methods

The program used is SRAC (Standard Reactor Analysis Code) 2K6 with JENDL 4.0 library [10][11]. The simulation was used to obtain the burn-up parameters of infinite multiplication factor (kinf), burn-up level, conversion ratio, and atomic density of U-235, U-238, and Pu-239. Nuclear fuel cells are pin-shaped with a cylindrical geometry, according to the design specifications in the GFR, as shown in Table 1. Also, this shape specification follows several existing studies.

Table 1. Fuel cell specifications

Parameter	Value/Description
Fuel (Fuel)	Uranium Metal (U-10%wtZr)
Cladding Material	Stainless Steel (SS316)
Coolant	Helium
Fuel cell geometry	Cylinder
Pitch pins	1.4 cm
Fuel volume fraction: cladding: coolant	55-65%: 10%: 25-35%
Burn-up time	100-160 years

3. Results and Discussion

Figure 1 shows the change in the value of the infinite multiplication factor (k_{inf}) with variations, without enrichment and enrichment at 2-10%. At 2-6% enrichment, the value of infinite multiplication factor (k_{inf}) is <1, while 8-10% enrichment showed a value of k_{inf} as >1. The high level of enrichment causes the density of U-235 to increase and impacts the number of neutrons produced, thereby causing the value of the infinite multiplication factor (k_{inf}) to be >1. This shows that the greater the enrichment of an element, the higher the value of the infinite multiplication factor (k_{inf}) obtained.

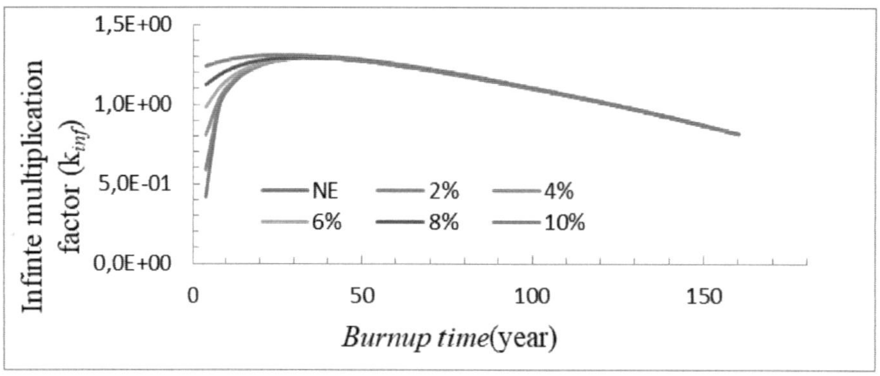

Figure 1. Infinite multiplication factor (k_{inf}) for 100,120, 140, and 160 years of burn-up

The burn-up level is the energy released per unit mass of fuel as a result of combustion. According to Figure 2, the burn-up level increases continuously over time. For example, in the year 100, 120, 140, and 160 the burn-up level was at 4.30E+05 MWd/Ton, 5.09E+05 MWd/Ton, 5.97E+05 MWd/Ton and 6.85E+05 MWd/Ton respectively. These values show that as the burn-up period increases, the amount of fuel burned also increases.

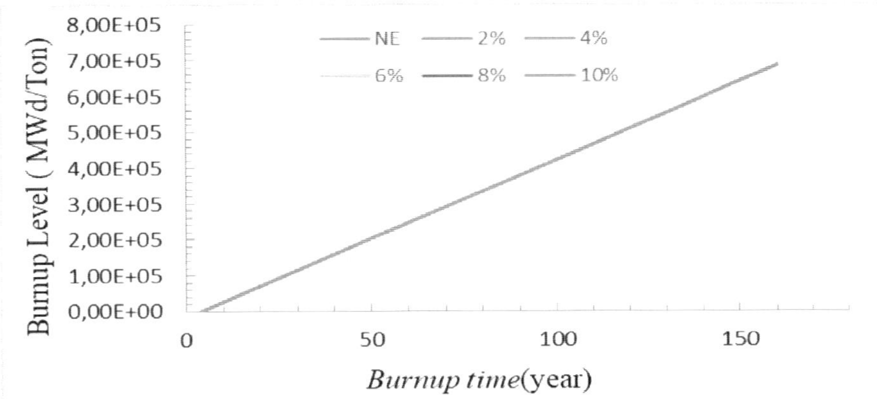

Figure 2. Burn-up levels for 100, 120, 140, and 160 years of burn-up time

The conversion ratio (CR) shows the ratio between the number of atoms of fissile fuel produced and the one consumed in the reactor. According to Figure 3, when the fissile atoms produced are more than the one consumed, the first ten years' conversion ratio for U-235 without enrichment experienced a very sharp decline, but afterward, it decreases slowly until the end of the burn-up time. Consequently, as the burn-up period increases, the fuel in the reactor runs out. Also, the enrichment level affects the value of conversion ratio obtained, hence the more the magnitude, the smaller the conversion ratio produced but the total density of U-235 is greater due to a large amount of enrichment.

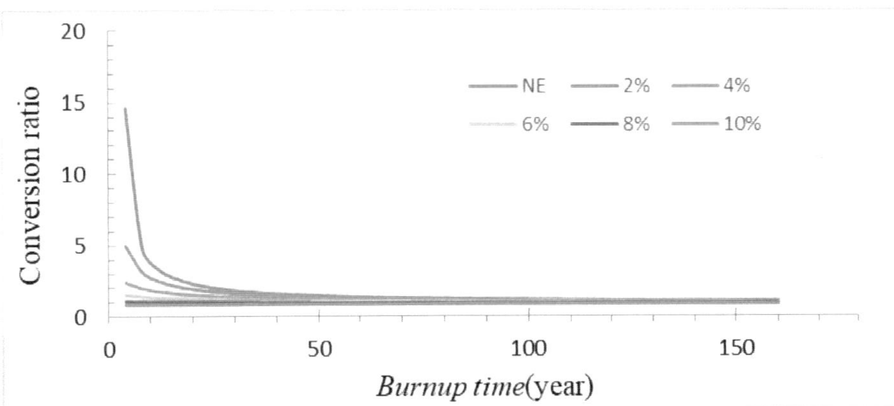

Figure 3. Conversion ratios for 100, 120, 140, and 160 years of burn-up

The atomic density distribution is used to obtain the population of the fission reaction in the fuel cell. Also, the fuel material in a nuclear reactor naturally undergoes a reaction while the device is operating. At a specific time, the fuel experiences a reduction in the amount due to the fission reaction and undergo nuclear transmutation. The primary materials used in this design are U-235 and U-238, which are transformed into other element such as Pu-239 during the reactions.

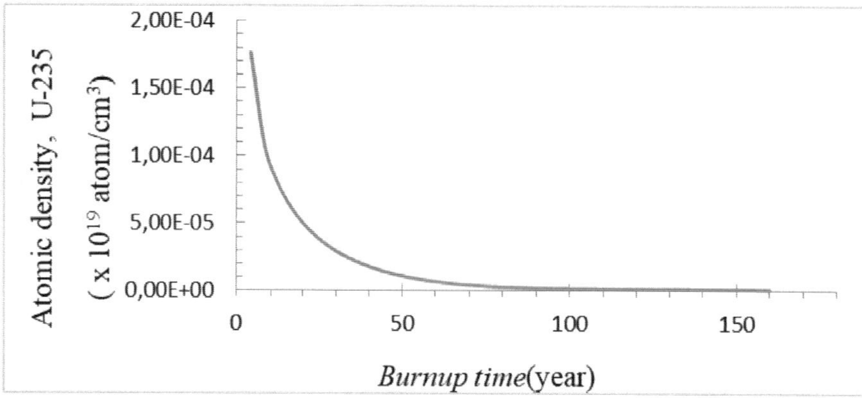

Figure 4. Distribution of atomic density of U-235 for 100,120, 140, and 160 years of burn-up

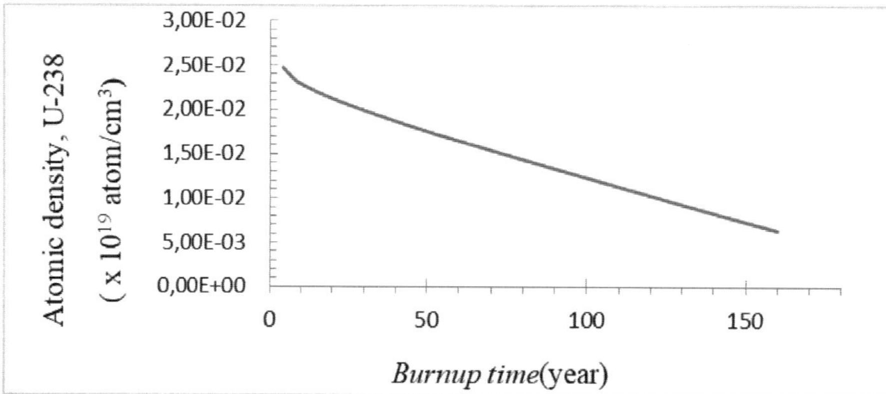

Figure 5. Distribution of atomic density of U-238 over 100, 120, 140, and 160 years of burn-up

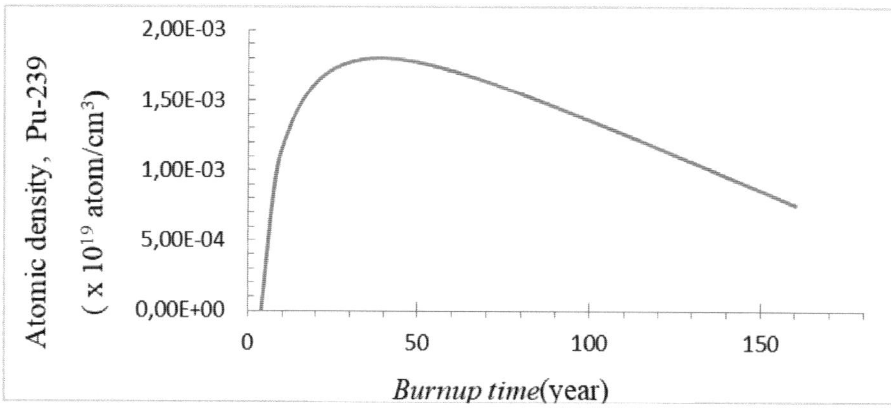

Figure 6. Distribution of the atomic density of Pu-239 over 100, 120, 140, and 160 years of burn-up

During the burn-up process, the atomic density of U-235 and U-238 changes to other elements due to the fission reaction in U-235 and the transmutation of U-238 which occurs in the fuel cell. Therefore, the atomic density of U-235 and U-238 decreases as the burn-up period increases, as shown in Figures 4 and 5. According to Figure 6, one of the resulting products, Pu-239 is absent because there is no fission reaction at the beginning, but when the burn-up process is running, it increases slowly and reaches its maximum value before decreasing significantly.

Basically, the fuel cell consists of three parts, which includes fuel, cladding, and coolant. The fuel functions in the ongoing fission reaction to produce the required heat, and the cladding determines the structural strength of the fuel cell to ensure there is no leakage, while the coolant anticipates the excess heat produced. The division of the volume fraction which is converted into 55%-65% fuel, 10% cladding, and 25%-35% coolant is carried out to ensure the comparison of the results obtained in each volume fraction division is changed.

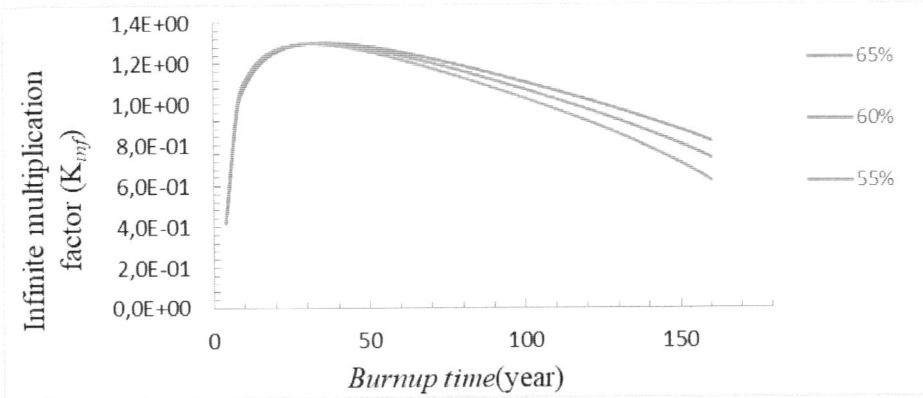

Figure 7. The infinite multiplication factor (k_{inf}) was compared by dividing the volume fraction 55-65% fuel, 10% cladding, and 25-35% coolant.

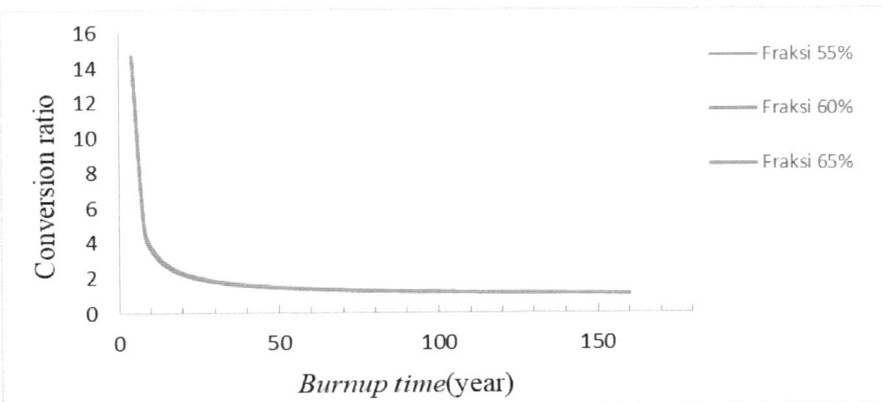

Figure 8. Comparison of the conversion ratio value by dividing the volume fraction 55-65% fuel, 10% cladding, and 25-35% coolant.

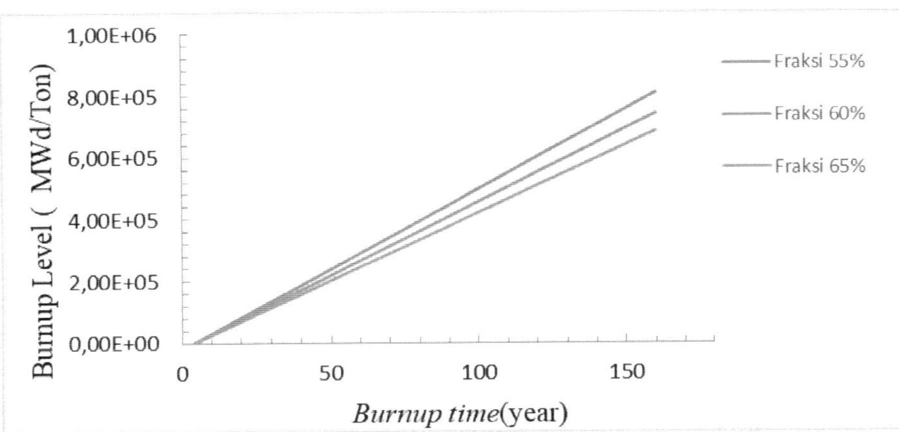

Figure 9. Comparison of burn-up level values by volume fraction distribution of 55-65% fuel, 10% cladding and 25-35% coolant.

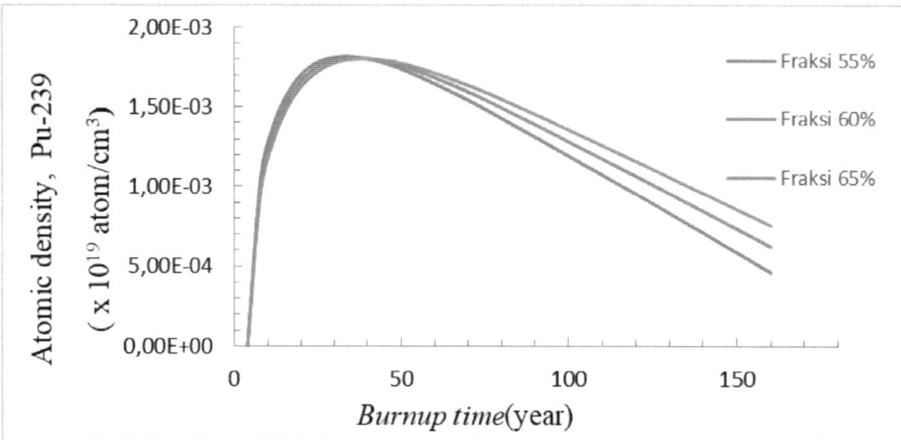

Figure 10. Comparison of the atomic density of Pu-239 with the volume fraction distribution of 55-65% fuel, 10% cladding, and 25-35% coolant.

According to Figure 7, the volume fraction of 65% fuel, 10% cladding, and 25% coolant obtains the highest infinite multiplication factor value in the thirty-sixth year of burn-up. However, for the volume fraction with 60% fuel, 10% cladding, and 30% coolant and the volume fraction of 55% fuel, 10% cladding, and 35% coolant, the multiplication factor value is smaller. This is because the large volume fraction of the fuel causes the fission reaction of the U-235 atom and the transmutation of the U-238 atom, as well as the capture of more neutrons.

Figure 8 shows no significant difference when the volume fraction is changed, hence the conversion ratio (CR) obtained is not affected. However, figure 9 shows that the value of the burn-up level obtained at the volume fraction of 55% fuel is greater than the one with 60% and 65%. This means that the smaller the volume fraction of fuel, the higher the burn-up level value obtained. Also, more fission reactions occur due to the small value of the fuel fraction used. According to Figure 10, the atomic density value of Pu-239 is more significant in the fuel volume fraction of 65% compared to the one of 55%-60%. Therefore, when the number of fuel volume fractions is large, the density of the number of Pu-239 atoms obtained is greater because more fission and neutron capture reactions occur.

4. Conclusion

The calculation of uranium metal fuel cell (U-10%wt Zr) with 4% enrichment during the burn-up time shows a critical value of kinf = 1,135. This increases continuously to 1,302 in the thirty-sixth year, before dropping towards the end of the burn-up time. Hence, a critical value is shown in each division of volume fraction comparison. However, the one with 65% fuel shows more stable results than the fuel volume fraction of 55% and 60%.

Acknowledgement

The publication of this article was funded by DIPA of Public Service Agency of Sriwijaya University 2021. SP DIPA-023.17.2.677515/2021, On November 23, 2020. In accordance with the Rector's Decree Number: 0010/UN9/SK.LP2M.PT/2021, ON April 28, 2021.

References

[1] Nurkholilah N and Fitriyani D 2019 Analisis Burn Up pada Reaktor Pembiak Cepat Berpendingin Pb-Bi dengan Variasi Fraksi Bahan Bakar dan Bahan Pendingin *J. Fis. Unand* **8** 184–90

[2] Alatas, Z., Hidayati, S., Akhadi, M., Purba, M., Purwadi, D., Ariyanto, S., Winarno, H., Rismiyanto, Sofyatiningrum, E., Hendriyanto, Widyastono, H. dan S 2016 *Buku Pintar Nuklir* (Jakarta: Batan Press)

[3] Adiwardojo, Ruslan and Parmanto, EM and Effendi E 2010 Mengenal Reaktor Nuklir dan Manfaatnya *Badan Tenaga Nukl. Nas.*
[4] Walter, A. E. and Reynolds A B 1981 *Fast Breeder Reactors* (New York: Pergamon Press)
[5] Haryani N and Fitriyani D 2013 Pengaruh Variasi Bahan Pendingin Jenis Logam Cair Terhadap Kinerja Termohidrolik Pada Reaktor Cepat **2** 190–4
[6] Andris D, Fitriyani D and Irka F H 2016 Optimasi Ukuran Teras Reaktor Cepat Berpendingin Gas dengan Uranium Alam sebagai Bahan Bakar **5** 21–7
[7] Cinantya; D and Fitriyani D 2014 Analisis Neutronik Pada Reaktor Cepat Dengan Variasi Bahan Bakar (UN-PuN, UC-PuC Dan MOX) *J. Fis. Unand* **3** 1–7
[8] Riska, Fitriyani D and Irka F 2016 Analisis Neutronik pada Gas Cooled Fast Reactor (GCFR) dengan Variasi Bahan Pendingin (He, CO2, N2) *J. Fis. Unand* **5** 28–34
[9] Monado F, Ariani M, Royani I and Su'Ud Z 2020 Comparative study of conceptual design of gas-cooled fast reactor core type tall versus pan cake based on MCANDLE-B burn up strategy *J. Phys. Conf. Ser.* **1568**
[10] Okumura K, Kugo T, Kaneko K and Tsuchihashi K 2007 *SRAC2006: A comprehensive neutronics calculation code system* (Japan)
[11] Susanty E, Ariani M, Su'Ud Z and Monado F 2019 Calculation of burnup fuel cell uranium metallic with carbon dioxide cooled *J. Phys. Conf. Ser.* **1282**

Liquefaction alarm prototype using arduino uno microcontroller

S Saehana[1] and A Lala[1*]

[1]Sub-Department of Physics Education, Faculty of Teacher Training and Education, Universitas Tadulako, Jl. Soekarno Hatta KM. 9 Kampus Bumi Tadulako Tondo Palu-Central Sulawesi, Indonesia

*Corresponding author: Andryaztikalala20@gmail.com

Abstract. This study aims to create an alarm tool that can detect disasters such as earthquakes and liquefaction in Palu City. It was laboratory research conducted in Sub department of Physics Education, Tadulako University. Research and experiments with modelling tools were conducted to simulate soil conditions during the earthquake and liquefaction. The research sample was focused on soil samples that have been affected by liquefaction disasters. The method used was the waterfall with the procedures of requirements analysis, system design, implementation, and testing of Arduino programs and software as ground movement monitoring. This study indicated that a liquefaction alarm prototype using a microcontroller component in the form of Arduino Uno, soil moisture sensor, and MPU6050 sensor could detect a natural liquefaction disaster was signed by a siren sound.

1. Introduction

Natural disasters are usually triggered by natural factors, non-natural factors, and human behavior. It may cause environmental damage and casualties [1]. Each country has different disaster vulnerability, including Indonesia. As located on three active plates and the equator, Indonesia has numerous active volcanoes and is highly prone to other natural disasters [2]. Only in a one-year period, thousands of disasters have occurred across Indonesia, including landslides, floods, earthquakes, tsunamis, and others. As of 2018, there have been 2,572 disasters that caused approximately 4,814 people lost their lives and were declared missing, 21,000 people were injured, and 300 houses were damaged [3].

Central Sulawesi is a province of Indonesia that has an active fault, called the Palu-Koro fault. On September 28, 2018, the Palu-Koro fault experienced a powerful movement, recorded on the seismographs of the earthquake measuring 7.4 on the Richter Scale [4]. In addition, the earthquake also triggered a tsunami as high as 2.2 meters that occurred in the coastal areas of Talise, Mamboro, and Donggala [4]. On the same day, there was a liquefaction disaster due to vibrations caused by an earthquake in several areas of Palu, namely Petobo, Balaroa, and Jonoge. It was the first time for the community around Palu to witness and experience such an event.

Liquefaction is an event that changes the structure and properties of the soil and the pressure in the soil due to movement, which causes the soil to be unable to control the strength and stiffness properties of the previous soil [5]. The 2011 Tohoku earthquake off the Pacific coast in Japan also caused liquefaction of large areas. In particular, severe liquefaction was reported in the northern part of the reclaimed land around Tokyo Bay [6]. Liquefaction has occurred in many locations as an impact of severe shaking. However, a local person in Central Sulawesi has not been familiar with the term liquefaction before the 2018 earthquake tragedy.

Based on the records within a century (1905–2005), there have been more than ten earthquakes with more than 4.5 SR in Palu City and its surroundings. Based on topographical, geological, and seismological conditions, the Palu City area can be damaged by earthquakes and secondary disasters (tsunami, liquefaction, and cliff landslides). On May 20, 1938, an earthquake with a magnitude of 7.6 on the Richter scale shook throughout the island of Sulawesi [7]. The liquefaction micro zonation mapping using the Ground Shear Strain (GSS) approach in Palu City and its surroundings shows that the potential liquefaction is in several villages: Baru, Besusu, Bayaoge, Nunu, Siranindi, Kamonji, Silae, Lare, Talise, and Kabonena [8].

Reviewing, studying, and analyzing all the supporting factors and the potential for such a large liquefaction disaster led to the importance of making an effort and action to prepare psychologically and physically. It is to reduce the risk and impact of the disaster, both before, during, and after the disaster occurs. However, liquefaction disasters due to earthquakes can come out of the blue, so it is necessary to have an early warning tool to inform the existence of liquefaction.

Research conducted by Ginting [9] produced an earthquake detector design tool equipped with an early warning alarm and automatic light control. It is one of the steps to accelerate evacuation in the event of an earthquake [9]. This tool was designed with a Hall Effect sensor A3144 as a vibration detector and a PIR sensor used to detect human body temperature.

An earthquake alarm system based on the AVR Atmega 16 microcontroller using a piezoelectric sensor shows an active alarm if the sensor output voltage is 1 volt and the sensor is still sensitive to detect vibrations up to 200-centimeter distance for a load drop height of 30 cm. The system consists of automation capable of activating the Mp3 player, activating an alarm when vibration is detected. This alarm will continue to be active if there is still a vibration, and this alarm will be deactivated after the vibration is no longer detected for 5 minutes. The earthquake alarm test was carried out through a straightforward process by dropping a load of 100 grams, 200 grams, and 300 grams. The falling distance to the sensor and the falling height of the load were 10 cm, 20 cm, and 30 cm. The reference used to indicate the occurrence of earthquake vibrations is 2 MMI (Modified Mercalli Intensity), as stated in research conducted by Rahman and Yusfi [10].

There has been no alarm that functions as an early warning of an earthquake and liquefaction disaster, especially in Palu City. Accordingly, the researchers were interested in making a liquefaction alarm prototype to detect earthquakes and liquefaction based on the Arduino Uno microcontroller. The product of this study can be used as a prototype of an alarm that can detect the liquefaction phenomenon as part of a form of natural disaster mitigation.

2. Methods

This research was a laboratory experiment implemented to create and produce a liquefaction alarm based on an Arduino Uno microcontroller. The research was carried out at the Physics Laboratory of the Faculty of Education, Tadulako University, Palu, Central Sulawesi, in September 2020. The liquefaction alarm prototype testing was carried out in a transparent container with a size of 30 x 23.5 cm with a soil sample of 2 kg and a volume of 1.5 L of water, obtained a water content of 75% with a dry volume weight of 0.591 gr/cm^3 and the wet volume weight of 0.946 gr/cm^3.The sample used was a soil sample from an area affected by liquefaction: the Petobo area, Palu City. This study conducted testing of the tool's string and the system as a whole. The data were collected from the analysis of the results provided by the tool.

3. Results and Discussion

3.1 Testing Purpose

The tests were carried out to test the tool's performance. In addition, as a program in the form of a system application, it is necessary to test the relationship between hardware and software. It was expected that the test could detect whether the designed tools and applications worked as desired or not. Initial testing was done by testing the components separately and then continued by testing the whole system.

3.2 Tool's String Test Results

The design was based on the liquefaction alarm prototype using the Arduino Uno microcontroller using two types of sensors, namely the soil moisture sensor and the MPU6050 sensor with gyroscope and accelerometer. The results of the tests carried out are as follows:

3.2.1 Soil moisture sensor display

The soil moisture sensor tests detected soil moisture on waterless soil (dry soil) and wet soil until it melted. The test results are shown in Table 1 and Figure 1.

Table 1. Testing the soil moisture sensor string

No	Test Type	Sample	Result
1	Waterless Soil	2 kg of soil	Generate sensor readings and display data on the application interface
2	Wet Soil	2 kg of soil + 1.5 L of water	Generate sensor readings and display data on the application interface

Figure 1. The string of soil moisture sensor test results

3.2.2 The display of sensor mpu6050 with gyroscope and accelerometer

The MPU6050 sensor test was carried out to detect ground movement by placing the sensor on the test container and then giving it a vibration through a drill considered an earthquake vibration. The test results are shown in Table 2 and Figure 2.

Table 2. MPU6050 sensor string testing

No	Test Type	Sample	Result
1.	Waterless Soil	2 kg of soil	Generate ground motion values and display waves on the application interface
2.	Wet Soil	2 kg of soil + 1.5 L of water	Generate ground motion values and display waves on the application interface

Figure 2. Test results of MPU6050 sensor's string

3.3 Overall System Test Results

System testing is executing a software system to determine whether the system matches the system specifications and runs as intended. This testing was intended to determine whether the software's functions, inputs, and outputs followed the required specifications [11]. This test was carried out by combining all sensor components along with the application software used.

Table 3. Overall system testing

No	Test Type	Sample	Result	
			Buzzer	Wave
1.	Waterless Soil	2 kg of soil	Short beep sound with 1-sec interval	Displays waveform reading
2.	Wet Soil	2 kg of soil + 1.5 L of water	Siren sound	Displays waveform reading

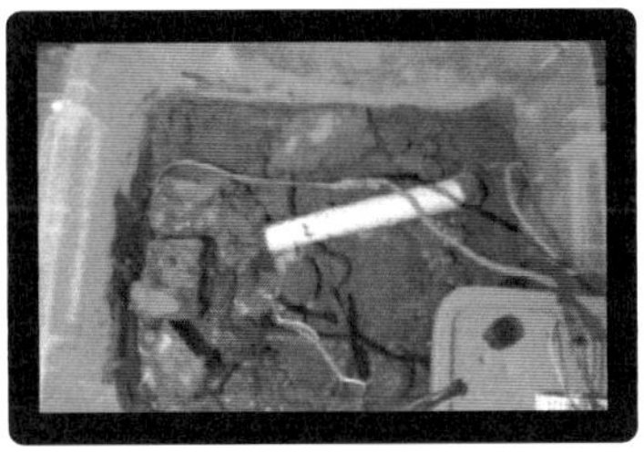

Figure 3. Overall test results

The working principle of the liquefaction alarm prototype in this study is that the alarm will sound when the pore water pressure in the soil continues to increase, causing the soil from a solid-state to become more liquid due to earthquake vibrations. In this liquefaction alarm prototype design and manufacture, each component has been tested both separately and as a whole. It resulted in accordance to what has been programmed. After testing a series of tools on the liquefaction alarm prototype, it was found that the mpU6050 sensor with gyroscope and accelerometer and soil moisture sensor functioned properly, as shown in the test results in Table 1 and Table 2. It shows that each sensor can display sensor reading data on the interface application. While in the overall system test, which can be seen in Table 3, on soil testing without water content, the buzzer sound produced was in the form of a tone with short beeps with an interval of 1 second. This pattern of alarm sound indicates the area is experiencing an earthquake. Meanwhile, when the sample soil was added water, the sounds become siren pattern. It warns that the liquefaction had occurred.

The liquefaction alarm prototype took place in a transparent container measuring 30 x 23.5 cm with a composition of 2 kg of soil and 1.5 L of water. The mixing results found that the water content in the liquefaction prototype was 75%, the dry volume weight was 0.591 gr/cm^3, and the wet volume weight was 0.946 gr/cm^3. In this study, the high-water content in the prototype caused the low bearing capacity of the soil, which resulted in the low ability of the soil to withstand maximum pressure or load, which then caused liquefaction. It is relevant to the reference [12], which states that the bearing capacity of the soil is obtained from its shear strength. It can resist shear stress when the soil is loaded. The water content significantly affects the soil type; adding or reducing water content up to 20% can reduce the value of shear strength which is quite significant and will affect land stability. The soil samples used were from areas affected by liquefaction, namely the Petobo area, Palu City. The type of soil in the area is poorly graded sand containing little or no refined grains, commonly referred to as non-plastic (non-cohesive) soil [13]. Poorly graded soil types tend to be unstable and more susceptible to liquefaction than well-graded soil types [13].

This prototype liquefaction alarm used two types of sensors: the mpu6050 sensor with accelerometer and gyroscope to detect ground movement. By using the accelerometer sensor, a ground movement detector can detect the occurrence of shifts and ground movements so that losses can be minimized [14]. Then, the soil moisture sensor was used to measure the increase in humidity level due to pressure on the soil pore water. Adding water with good conductor properties can result in a small resistance reading and vice versa. If the water content in the soil is reduced or low, it produces a considerable resistance value [15] so that it can be used to detect soil liquefaction due to vibration. This liquefaction alarm device has produced a tool to warn of natural disasters (earthquakes and small-scale liquefaction) and monitor earthquakes' movement.

In this prototype liquefaction alarm system, it is limited to several factors causing liquefaction. Therefore, there are still weaknesses. The amount of energy given to produce earthquakes has not become a fixed variable, the type of soil used has not been filtered in detail, the range of tools was still small, and the monitoring system could not be accessed online. Several factors limit this prototype liquefaction alarm because it requires further study and a longer time in the research process.

The liquefaction alarm is also expected to be a medium for learning physics in classroom learning so as to complement the existing science instruments [16], energy conversion tools [17] and fluorine-doped tin oxide [18].

4. Conclusion

An alarm prototype using a microcontroller component in the form of Arduino Uno to detect earthquakes and liquefaction has been developed. Although it was still in a form of prototype, it can be used in the community as the early warnings for natural disasters and monitor the phenomenon of liquefaction. Researchers suggest that this tool is developed more digitally, like using electromagnetic waves, so that the intended reach is broader compared to this study. In addition, the use of frequency on the vibrator set is recommended to detect the amount of energy generated by the vibration. Also, researchers recommend using soil types with various samples of soil grains with different sizes and screening numbers and an online and android-based warning system so that all the public can know it.

Acknowledgement

We thank kindly Hibah Penelitian Terapan Kompetitif Nasional DIKTI and Program Kompetisi Kampus Merdeka (PK-KM) of Ministry of Education, Culture, Research and Technology of The Republic of Indonesia, under the funding 302/E4.1/AK.04/PT/2021 and 119/E1/KM.05.03/2021, respectively.

References

[1] Anon 2003 Undang-Undang Republik Indonesia Nomor 24 Tahun 2007 Penanggulangan Bencana *Lembaran Negara Republik Indones. Tahun*

[2] Yanuarto T, Pinuji S, Utomo A C and Satrio I T 2017 *No Title* (Jakarta Timur: Pusat Data Informasi dan Humas BNPB)

[3] Rosyida A, Nurmasari R and Suprapto S 2019 Analisis perbandingan dampak kejadian bencana hidrometeorologi dan geologi di indonesia dilihat dari jumlah korban dan kerusakan (Studi: Data kejadian bencana indonesia 2018 *J. Dialog Penanggulangan Bencana* **10** 1–104

[4] Nasional T P S G 2018 Kajian Gempa Palu Sulawesi Tengah 28 September 2018

[5] Kusuma R I, Mina E and Kurniawan N 2014 No Title *J. Tek. Sipil* **3** 73–82

[6] Fukutake K and Jang J 2013 Studies on soil liquefaction and settlement in the urayasu district using effective stress analyses for the 2011 off the Pacific Coast of Tohoku Earthquake *J. JSCE* **1** 307–321

[7] Ramadhani S 2011 Kondisi seismisitas dan dampaknya untuk kota palu *J. Tek. Sipil Dan Infrastruktur* **1**

[8] Kurniawan C W 2019 *No Title* (Yogyakarta: UNIV Pembangunan Nasional “Veteran”)

[9] Ginting W G 2021 *No Title* (UNIV Sumatera Utara Medan)

[10] Rahman, M N and Yusfi M 2015 Rancang bangun sistem alarm gempa bumi berbasis mikrokontroler avr atmega 16 menggunakan sensor piezoelektrik *J. Fis. Unand* **4** 350–357

[11] Sukamto R A and Salahuddin M 2013 *Rekayasa perangkat lunak terstruktur dan berorientasi objek* (Bandung: Informatika)

[12] Agustina D H and Elfrida M 2019 Pengaruh perubahan kadar air terhadap kekuatan geser tanah lempung *Sigma Tek.* **2** 115–122
[13] Pratama A 2020 *No Title* (Univ Negeri Semarang)
[14] Hidayat M T 2018 *No Title* (UNIV Siliwangi)
[15] Jupri A and Muid A 2017 Rancang bangun alat ukur suhu, kelembaban, dan ph pada tanah berbasis mikrokontroler ATMega328P *JEPIN (Jurnal Edukasi dan Penelit. Inform.* **3** 76–81
[16] Saehana S, Ali M and Supriyatman S 2019 *J. Pendidik. IPA Indones.* **8** 241–246
[17] Rohman A, Komang Werdhiana I and Saehana S 2021 *J. Phys. Conf. Ser.* **1760**
[18] Izzah A and Saehana S 2021 *J. Phys. Conf. Ser.* **1760**

Analysis of fault patters based on erarthquake data on the land of Sumatera island

A Basid[1*] and B Munir[2]

[1]Department of Physics, Faculty of Science and Technology, UIN Maulana Malik Ibrahim Malang, Indonesia
[2]Observer of geological disasters

faizabasid@fis.uin-malang.ac.id

Abstract. Geodynamics of Sumatra is interesting to study because of the unique geological setting Geodynamics of Sumatra is interesting to study because of the unique geological setting and high seismicity. This high seismicity is caused by the many faults found on land and in the surrounding waters. This paper presents the results of research that aims to determine fault patterns both on land and in the waters around Sumatra based on earthquake data from 1960-2000. The area under study is at the coordinates of 6°N - 6°S and 95°E - 109°E at an epicenter depth of <60 km with a magnitude between 4-10. The area is divided into two zones, namely the front arc and the Barisan arc. Data were obtained from ISC and Global CMT. By using the Focal Mechanism Method, the results show that the fault pattern in the Sumatran forearc Basin zone is dominated by an Reverse Fault located in the accretion zone while on the mainland of Sumatra it is dominated by Strike Slip along the Sumatran Fault System. By knowing the position and pattern of the fault, especially on the mainland of the island of Sumatra, it can be used as a reference for spatial planning. In addition, further studies will also be able to learn about the dangers or disasters caused by the fault pattern.

1. Introduction

The accumulation of natural disasters in Indonesia from the beginning of the year to the end of August 2021 was recorded as 1,805 incidents. Wet hydrometeorological disasters dominate the occurrence of this disaster. Especially for earthquakes, 23 destructive earthquakes were recorded [1]. For earthquake disasters during January to July 2021, the Meteorology, Climatology and Geophysics Agency (BMKG) recorded 4,701 earthquakes in Indonesia from which they could not be felt to those that were destructive. The results of BMKG monitoring show that the trend of earthquakes in Indonesia will increase in 2021, reaching 300 to 400 earthquakes every month [2]. The existence of the Indonesian archipelago which is located in an active tectonic plate boundary zone is the main cause of earthquakes that occur almost every day. One of the large islands in Indonesia which is located on the boundary of the world's great plates is the island of Sumatera. This geological position causes the island's seismicity to be quite high.

Sumatera geodynamics is interesting to study, especially its high level of seismicity. Earthquakes that occurred in North Sumatera and Aceh throughout the month of May 2021 alone were recorded at the Regional Earthquake Center (PGR) I Center for Meteorology, Climatology and Geophysics (BBMKG) Region 1 Medan as many as 407 events [3]. This high seismicity is due to the fact that this island is one of the traces of the Indonesian continent which is formed from broken, united and uplifted continental chunks originating from Gondwana and Cathaysian [4]. Almost all of the microscopic continents in Indonesia are surrounded by active plate margins that form a complex plate meeting path of the world's three major plates and nine minor plates that meet each other [5]. This geological arrangement causes quite high seismicity in almost all islands in Indonesia due to the dynamics and interactions between these plates. The dynamics and interactions between these plates also cause the formation of derivative tectonic structures such as faults, local faults, folds, land subsidence in almost all of the Indonesian archipelago[6].

Earthquakes occur due to the sudden release of energy when the rock at the source location is unable to withstand the forces due to the relative motion between plates [7]. This energy propagates in the form of seismic waves that move towards the surface and cause shocks or vibrations [8]. One of the mechanisms or methods for analyzing the waveforms generated by earthquakes recorded by a number of seismographs is the Focal Mechanism Solution (FMS). The complete characterization of an earthquake's focal mechanism provides important information, including the origin time, epicenter location, focal depth, seismic moment and the magnitude and spatial orientation of the 9 components of the moment tensor. All of this information can be obtained in seconds by structural geologists who know how to interpret FMS graphic depictions, known as "beach ball" diagrams [9].

Beachball diagram or also known as Focal Ball has a three-dimensional shape and is analyzed for the direction of movement of the origin of the P wave. Visual interpretation is done by dividing the focus ball into two parts, the top and bottom. Then the lower half of the sphere is irradiated from below, so that a two-dimensional image of the sphere is obtained, which is called the focal sphere diagram [10]. From this diagram it will be possible to know the parameters of the orientation of the fault plane of the fault that caused the earthquake. The fault parameters consist of strike (φ), dip (slope) (δ), rake (slip) (λ). Strike is a fault plane move measured from north to east (0° - 360°). Dip is the angle formed by the fault plane with the horizontal plane, and is measured in a vertical plane whose direction is perpendicular to the fault direction ($0^o \leq 90^o$). Rake or slip, is the hanging-wall movement angle with respect to strike ($-180^o \leq 180^o$), rake is positive for ascending fault and negative for descending fault [11].

Earthquake waves propagate in all directions in the form of various wave phases with P waves as the phase with the greatest speed. This direction of motion is the first to be observed to study the source mechanism. Studying the P wave mechanism is not only to study the initial motion of the earthquake and the solution of the fault plane, it can also be used to determine the motion of tectonic plates which is used to determine the relative motion of the lithosphere [12]. Fault is a fracture that has a clear shift. The shifts range from a few millimeters to hundreds of meters while their lengths range from a few decimeters to thousands of meters. Faults can occur in all kinds of rocks with various types. Faults in rock structures can lead to topographic changes and developments, change the flow of water below and above the surface and damage rock stratigraphy and so on [13].

This studi of focal mechanism is carried out in disaster mitigation efforts because earthquakes are difficult to predict when, where and how large they are accurately. This disaster must continue to be watched out for because it often causes damage to infrastructure and even casualties. Mitigation is carried out not to reduce or prevent disasters, but to reduce their impact, both on infrastructure and on casualties. Mitigation efforts can be carried out through mapping the affected area and increasing the capacity of the affected community through awareness and concern in dealing with disasters. In this study, apart from knowing the types of faults that cause earthquakes, the fault parameters will also be known. From the magnitude of the earthquake, further studies will be able to map the level of risk of damage from the earthquake. By knowing the types of faults and their parameters, it will also be possible to study the spatial arrangement around the fault and the development of its infrastructure that is adaptive to the characteristics of the fault. The most important thing from the results of this study is to know the points or locations and types of faults on the mainland of the island of Sumatra more accurately so that the impact can be known if an earthquake occurs and mitigation is carried out continuously.

2. Method

To determine the types of faults that cause earthquakes in Sumatera and its surroundings, this paper uses the Focal Mechanism Solution or Focal Mechanism method. This method is used to interpret a fault form in the form of a focal sphere caused by the initial motion of the P wave from an earthquake. From the focus ball, information on the shape of the fault and the direction of the slip will be obtained. The parameters of the fault that can be studied from this method include strike, dip and slip.

The data processing and analysis of this study was carried out in three stages, namely processing seismicity data with Magnitude, Focal Mechanism using data from ISC and CMT global catalog and geological mapping and focal data using Arcview GIS 3.3 software. Earthquake data obtained from the ISC and the CMT global catalog in determining the faults are classified according to the desired depth, which in this study uses earthquake data with a depth of less than 60 kilometers (shallow earthquake) with geographical limitations 6°N - 6°S and 95°E - 109°E and using magnitude 4-10. The classification of the data that has been obtained is then carried out by fault analysis based on the obtained focal ball (beachball) data.

There are three zones analyzed for fault types, namely the Forearc Basin zone, Sumateran Fault zone and Backarc Basin zone. Earthquake data from these three zones is already available and the analysis and interpretation is adjusted to the geological data. Due to the less significant seismicity in the Backarc Basin zone, it was not reported in this paper.

3. Result and Discussions

3.1. Distribution of seismicity on the island of Sumatera

Figure 1 shows the distribution of seismicity and its magnitude between 1960 and 2000 in Sumatera at the geographical boundary between 6°N - 6°S and 95°E - 109°E. From this distribution, it can be seen that the earthquakes in Sumatera mostly occurred in the west of the island and along the Sumatera fault area. The seismicity in the western region of Sumatera is along the acceleration zone or in the subduction zone where one plate is subducting under the other. In this zone the Indian Ocean plate is subducting under the Eurasian plate and this strip is called the Benioff Strip. On the mainland of the island of Sumatera, many earthquakes occur along the Sumatera Fault where this fault is one of the land faults in Indonesia which is very dynamic so earthquakes often occur along this fault.

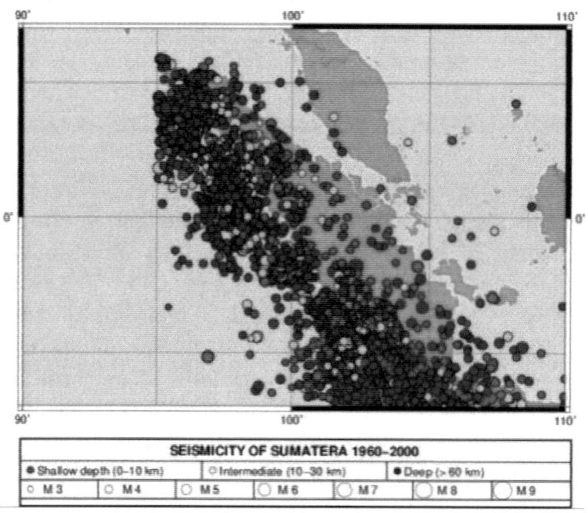

Figure 1. Distribution of seismicity of Sumatera.

If a transverse incision is made as shown in Figure 2, it will show the distribution of seismicity in the vertical plane under the incision as shown in Figure 3. From Figure 3, it can be seen that in one plane in the vertical direction there have been many earthquakes with varying depths and magnetoes. In Figure 3 at a depth between 30-50 km there is an earthquake phenomenon with a relatively the same depth as

Journal of Physics: Conference Series **2126** (2021) 012003 doi:10.1088/1742-6596/2126/1/012003

if forming a boundary pattern. It is therefore necessary to further study whether this is a weak zone that might show plate boundaries, or whether the Sunda Microplate is shifting over the Eurasian Plate.

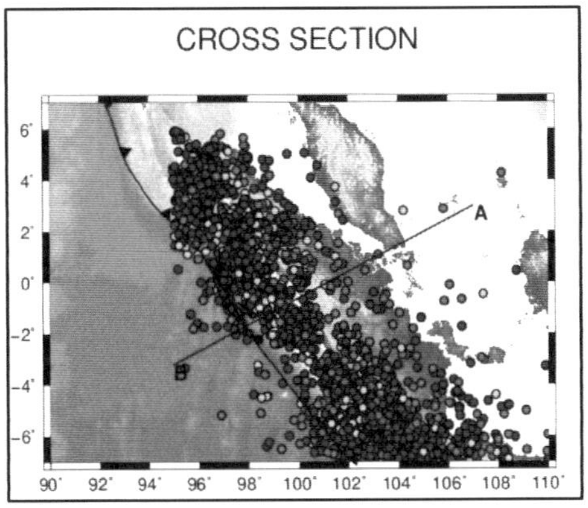

Figure 2. Cross section on seismicity

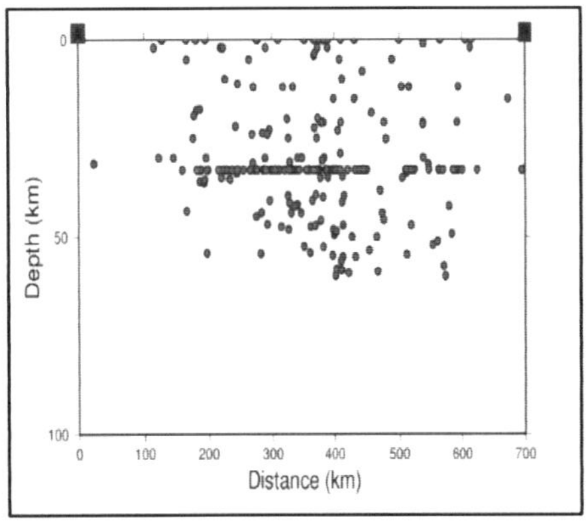

Figure 3. Distribution of seismicity in the vertical plane under the cross section

3.2. Types of faults on the island of Sumatera and its surroundings

By using the focal mechanism method to analyze the types of faults on the Sumateran fault island and its surroundings, four types of faults were obtained, namely strike slip, reverse fault, normal fault and oblique. Table 1 shows the types of the 24 faults that exist along the Barisan Zone. In this zone, it can be seen that the strike slip type or shear fault dominates in this zone.

Table 1. Types of faults along the Barisan zone

Longitude (°)	Beachball	Fault Type	The amount
95-105 EL	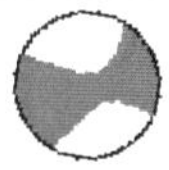	Strike Slip	15
		Reverse Fault	3
	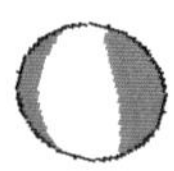	Normal Fault	1
	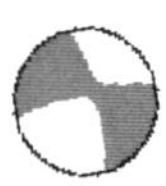	Oblique	5

In the Forearc Basin Zone, the types of faults that caused 153 earthquakes occurred in this zone were studied. Table 2 shows the types of faults that caused 153 earthquakes in the Forward Arc zone. From this table, it can be seen that the reverse fault type dominates in this zone with 68 occurrences. This shows that the dynamics in the subduction zone are the cause of most of the earthquakes in this zone.

Table 2. Types of faults along the Forearch Basin zone

Longitude (°)	Beachball	Fault Type	The amount
	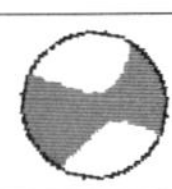	Strike Slip	32

95^0-105^0 EL		Reverse Fault	68
	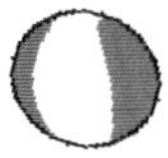	Normal Fault	20
		Oblique	33

3.3 Geology and Tectonic Studies

The geodynamics and tectonics of the island of Sumatera in particular and the entire archipelago in Indonesia in general are strongly influenced by the dynamics of the three mayor world plates. The three major plates are the Indian-Australian Ocean Plate moves from south to north, the Pacific Plate from east to west and the Asian Plate moves from north to south southeast [14]. Based on the data on the types of faults that cause earthquakes in Sumatera as stated above, it can be said that the seismicity on this island is strongly influenced by the dynamics of the subduction zone in the western zone of the island. This zone is a convergent plate boundary that extends from north to south and extends to the south of the island of Java. The most intense tectonic plate interactions occur in the western and southern parts of the Indonesian archipelago and form the Sunda Arc [15]. This Sunda arc is the meeting between the Indian-Australian plate and the Eurasian Plate. The movement of plates in this zone also causes the formation of faults, folds, land subsidence and so on. Other secondary processes due to activity in this subduction zone have also led to the formation of troughs, non-volcanic outer arcs, forearcs, volcanic arcs, and back arc basins [16].

The dynamics of this zone also causes the formation of regional geological structures on the Sumatera mainland, among which the best known is the Sumatera Fault System. Based on the results of mapping the types of faults as in table 1 shows that in the Sumatra Fault System with a reverse fault. The results of mapping the types of faults also showed that earthquakes in the path of the Sumatra fault system were less than in the face arc. Bellier et al in this case also said that earthquakes with magnetudes of 5 or more or less occurred on the Sumatran mainland and most earthquakes occur in the forearc basin [17]. Due to subduction in the forearc basin zone also causes the formation of a line volcanoes known as Bukit Barisan.

As stated in the previous section, the geological and geodynamic structure of Sumatra is indeed quite unique. In terms of seismicity, the fairly high seismicity on this island (figure 1) shows that the source of the earthquake on this island is caused by two causes, namely the foreach basin and the Sumatran Fault System. Of course, mitigation must be different because the location of the epicenter is different. Mitigation carried out for earthquakes whose epicenter is on land must be more careful and accurate because the source of the earthquake can be directly located in residential areas. The damage to infrastructure and the resulting loss of life will be more than that of an earthquake whose epicenter is in the water, even though the magnitude is smaller. Mitigation for the type of fault must also be different according to the type of fault and the parameters of the fault. For this reason, further studies on the appropriate mitigation for each fault are needed.

3. Conclusions

From the analysis of 177 earthquake data with a depth of <60 km that occurred in Sumatra and surrounding areas it can be concluded that the types of faults are strike slip, reverse fault, normal fault and obliques. In the forearc basin zone the type of fault is dominated by an reverse fault. From this type of fault shows that in this zone is an accrasi zone that is tectonically a subdction zone between the Indian-Australian Plate and the Eurasian Plate. In the zone along the Sumatra Fault System type of fault is dominated by a strike slip. The dominance of this type indicates that the land of the central part of

Sumatra fault is controlled by the Oblique Collision of the two plates proposed earlier. On the backarc basin there is no significant earthquake data so it is not studied in this paper.

Acknowledgments

Thanks to my colleague Amalia Hidayati, head and for all staffs of BMKG and thanks for all lecturers and staff in Physics Department of UIN Maulana Malik Ibrahim of Malang.

References

[1] BNPB 2021 Reflection of natural disasters throughout august 2021, The dominance of two different natural phenomena that affect preparedness efforts, https://bnpb.go.id/berita/refleksi-bencana-alam-sepanjang-agustus-2021-dominasi-dua-fenomena-alam-berbeda-yang-pengaruhi-upaya-kesiapsiagaan- (accessed: September 3, 2021 06:34 A.M)
[2] BMKG 2021 807 Earthquakes occur in indonesia during April 2021, https://twitter.com/ info BMKG/status/1388857141591613444 (accessed: August 21, 2021 8:23 A.M).
[3] Juariadi J 2021 Aceh and Sumatera were rocked by 407 earthquakes during May 2021, https://aceh. antaranews.com/berita/218770/aceh-dan-sumatera-diguncang-407-kali-gempa-selama-mei-2021 (accessed: August 21, 2021 08:34 A.M)
[4] Hutchison C S 1989 *Geological evolution of south-east asia oxford monographs on geology and geophysics no. 13. xv + 368 pp.* (Oxford: Clarendon Press)
[5] Bird P 2003 *Geochemistry Geophys. Geosystems* **4** 1027
[6] Milsom J 2003 *Field geophysics (the Geological Field Guide Series) edition III* (London: John Willey & Sons Ltd)
[7] Sunarjo S 2012 *Earthquakes popular edition* (Jakarta: Central BMKG)
[8] Helmstetter A, Sornette D and Grasso J R 2003 *J. Geophys. Res.* **108**
[9] Cronin V 2004 *A draft primer on focal mechanism solution for geologists texas* (Baylor University)
[10] Prasetya T 2006 *Earthquake* (Yogyakarta: Gitanagari)
[11] Keliti A and Paul G R 1980 *Quantitative seismology and methods* (San Fransisco: W. H. Freeman and company)
[12] Susilawati S 2015 Earthquake seismic wave propagation in studying the internal structure of the earth, http://repository.usu.ac.id/handle/123456789/18904 (accessed: August 22, 2021 06:34 A.M).
[13] Tjia H D 1997 *Geol. Indones.* **1** 13–27
[14] Indonesian Earthquake Map Revision Team 2010 *Summary of study results of the 2010 indonesia earthquake map revision team* (Bandung: Department of Public Works)
[15] Harjono H 2017 *Seismotectonic Sunda Arc* (Jakarta: LIPI Press)
[16] *Disaster Risk Study Preparation Team 2018 Indonesian Disaster Risk* (Jakarta: BNPB)
[17] Bellier R, Sebrier M, Pramumojoyo S, Beudouin T, Harjono H, Bahar I and Forni D O 1997 *J. of Geodin.* **24** 169–183

Journal of Physics: Conference Series **2126** (2021) 012004 doi:10.1088/1742-6596/2126/1/012004

DSSC with PEDOT-Carrageenan Electrolyte as Learning Media for Photovoltaic Concept Physics

N Sari [1], A J Palamba[1] S Saehana[1] and A W M Diah[2*]

[1]Sub-Department of Physics Education, Faculty of Teacher Training and Education, Universitas Tadulako, Jl. Soekarno Hatta KM. 9 Kampus Bumi Tadulako Tondo Palu-Central Sulawesi, Indonesia.
[2]Sub-Department of Chemistry Education, Faculty of Teacher Training and Education, Universitas Tadulako, Jl. Soekarno Hatta KM. 9 Kampus Bumi Tadulako Tondo Palu-Central Sulawesi, Indonesia.

*Corresponding author: anangwmdiah@gmail.com

Abstract. This study aims to create a DSSC using PEDOT-Carrageenan-based electrolyte as a learning media for photovoltaic physics concepts. This research was experimental research conducted in the laboratory of subdepartment of physics education, Tadulako University. In this study, DSSC **making** process used a substrate made from TCO-FTO polymer. The TiO_2 photoelectrode layer was deposited using the doctor blade technique and Ruthenizer as a dye-sensitizer and the platinum layer was deposited on the counter electrode using the sputtering technique. Process parameters that affect the characteristics of solar cells were analyzed and measured using the I-V characteristic curve. The samples measured consisted of several types of PEDOT-Carrageenan electrolytes, including Kappa, Lambda and Iota. The results of voltage measurements of each sample of the DSSC prototype were obtained quite well at a voltage of 91.8 mV using PEDOT-Carrageenan Kappa 1:2 electrolyte and the short circuit current obtained reached 57.6 uA. The best solar cell prototype produces an efficiency of 0.0003333728 %. It indicates that DSSC can be made by PEDOT-Carragenan electrolyte and can be used as a learning tool for photovoltaic physics concepts.

1. Introduction

The world's economic growth is proportional with the increasing demand of energy to meet the economic growth and human needs. However, the majority of the energy supply still depends on fossil fuels, such as coal, oil and gas. Since these resources are non-renewable natural resources, most nations have urged to use of renewable energy resources which are more environmentally friendly. In Indonesia for example, through the presidential decree of the Republic of Indonesia No. 5/2006 about the National Energy policy, the government encourages the use of renewable energy resources up to 5% of the total primary energy at the year 2025 [1]. Some renewable energy sources that frequently used as alternative energy sources to overcome the current energy crisis are solar, biomass, wind and hydropower energy resources.

The alternative energy resources are urgently required since the availability of fossil fuels as the main energy resource is running low. As a country that passed by the equator and has tropical climate, sun is one of the abundant and environmentally friendly alternative energy sources. Sunlight can be converted into electrical energy using solar cells by converting sunlight directly into electrical energy. Solar cells based on silicon technology, which is the result of the development of electronic semiconductor technology, is the most frequently used today [2]. Despite its popularity, its drawback was associated with the high cost of producing silicon that makes its consumption cost more expensive than fossil energy sources. In addition, the use of hazardous chemicals in the fabrication process contributes to another disadvantage of silicon solar cells. Therefore, another alternative has been found, namely Dye Sensitized Solar Cells (DSSC) [3].

DSSC is a power plant capable of converting sunlight into electric current [4]. DSSC attracts a lot of attention from academia and industry because it offers renewable energy solutions with low pollution levels, environmental friendliness and low cost of electricity production. DSSC consists of a photoanode made of Transparent Conductive Oxide (TCO) glass with a substrate (ITO) coated with a semiconductor such as TiO2, ZnO, SnO2, Nb2O5 with various methods such as doctor blade, screen printing, electroposition, and spin coating [5].

Mostly the dyes used as sensitizers can be synthetic dyes or natural dyes. Due to the toxicological and ecological reasons, the use of natural dyes is preferable and suggested than the synthetic dyes. Moreover, natural dyes are relatively cheaper and are more environmentally friendly. However, these products have disadvantages, including the different and inconsistent quality and quantity of the raw materials since they are influenced by the condition of the nature. Substances such as chlorophyll, beta carotene, anthocyanins, tannins, curcumin in plants can be used as sensitizers [6]. In recent years, many studies have used plants as natural dyes. These natural dyes are extracted from trees, flowers, fruits, and vegetables as absorbent substances (DSSC) [7][8][9][10][11].

Numerous research related to DSSC have been conducted. Most of them focused on increasing the energy performance value that was produced by DSSC and the materials development that can reduce the production costs [12]. Some research investigated the use of natural dyes using anthocyanin extract content, the use of pencil-type carbon catalysts and combustion products, and the use of iodine types [12]. A study of Camacho [13] about the use of electrolyte polymers in the form of a mixture of Poly (3,4-ethylenedioxythiophene) PEDOT Carrageenan polymers for DSSC found that an efficiency value was 0.421%. The value obtained indicates that there is potential for the development of an electrolyte polymer based on a carrageenan PEDOT mixture which is referred to a PEDOT/Carr polyblend.

This study aims to make DSSC type solar cells using PEDOT Carrageenan electrolyte and uses it as learning media for photovoltaic physics concepts. A research topic regarding the use of PEDOT Carrageenan mixture is still rarely carried out, especially on its use as a substitute for iodine electrolyte in the DSSC layer. On that account, this research can give new information and reference regarding the use of a mixture of PEDOT Carrageenan as an electrolyte in DSSC. In this study, researchers tested DSSC based on TiO2 nanoparticles using a synthetic *ruthinizer* produced by Solaronix and modified the use of iodine electrolyte using a mixture of PEDOT Carrageenan. In addition, the product of this study was intended to be used as a learning media. DSSC learning media is an active element that can convert sunlight into electrical energy, which is made of a wedge of semiconductor material with a positive pole and a negative pole. A simple DSSC that can be easily implemented in classroom learning has rarely been conducted. In previous studies, most focused on developing teaching materials in the form of written learning resources such as practicum modules or worksheets. Since the application of media in learning can increase the efficiency and activeness of the learning process [14][15], the availability of final product can be used to convey information related to the photovoltaic concept.

2. Methods

This research was laboratory experimental research. This research was conducted at the subdepartment of Physics Education Laboratory of Faculty of Teacher Training and Education, Tadulako University in August 2020. The samples used in this study were PEDOT Carrageenan Kappa 1:2, PEDOT Carrageenan Iota 1:1, Iota 1:2, Iota 1:4, PEDOT Carrageenan Lambda 1:1 and Lambda 1:2.

At the beginning, a study was conducted on the Fabrication of Dye Sensitized Solar Cell (DSSC) sing Ruthenizer as a synthetic dye. It began with the study of the characteristics of the DSSC and fwas followed by the characteristic tests of the current and voltage in the circuit with an I-V meter. The data were collected using a digital multimeter connected in series to the DSSC circuit. The data were analysed to measure the DSSC efficiency value.

Journal of Physics: Conference Series **2126** (2021) 012004 doi:10.1088/1742-6596/2126/1/012004

3. Results and Discussion

3.1. DSSC Solar Cell Characteristics

The DSSC prototype that has been made can be seen in Figure 1.

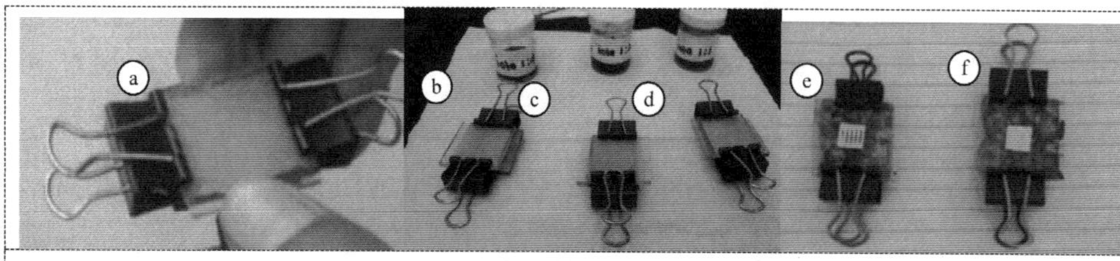

Figure 1. Sample of a DSSC prototype with different electrolytes; (a) Sample DSSC with PEDOT-Carrageenan Kappa 1:2 electrolyte, (b) Sample DSSC with PEDOT-Carrageenan Iota 1:2,1(c) Sample DSSC with PEDOT-Carrageenan Iota 1:2 electrolyte, (d) Sample DSSC with PEDOT-Carrageenan Lamda 1:4, (e) Sample DSSC with PEDOT-Carrageenan Lamda 1:1 electrolyte, and (f) Sample DSSC with PEDOT-Carrageenan Lamda 1:2 electrolyte

The DSSC prototype is assembled with a sandwich arrangement consisting of a working electrode (anode), electrolyte and counter electrode (cathode). The working electrode consists of TCO-FTO, TiO2 coating and dye. The counter electrode consists of a layer of Carbon (Platina) and TCO-FTO. The size of the working area on the TCO (working electrode) glass of each DSSC sample is different. For the DSSC sample with PEDOT Carrageenan Kappa 1:2 electrolyte, it has an area of 2.2 cm x 2 cm. For DSSC samples with PEDOT Carrageenan electrolyte Iota 1:1 has an area of 2.4 cm x 2.1 cm, Iota 1:2 has an area of 2.3 cm x 2.1 cm, Iota 1:4 has an area of 2.2 cm x 2.2 cm. For the DSSC sample with PDOT-Carrageenan Lamda 1:1 electrolyte which has an area of 2.5 cm x 2.8 cm, Lambda 1:2 has an area of 2.4 cm x 2.6 cm. Based on the six samples of the DSSC prototype built, each of them has a different size of TCO glass so that the size of the working area on the TCO glass is also different.

3.2. Current and Voltage Characteristics of Solar Cells with Circuit

To determine the performance of solar cells, measurements of current-voltage (I-V) characteristics were carried out in illuminated conditions by using a solar light source and by adjusting the resistance used. The circuit is shown in Figure 2.

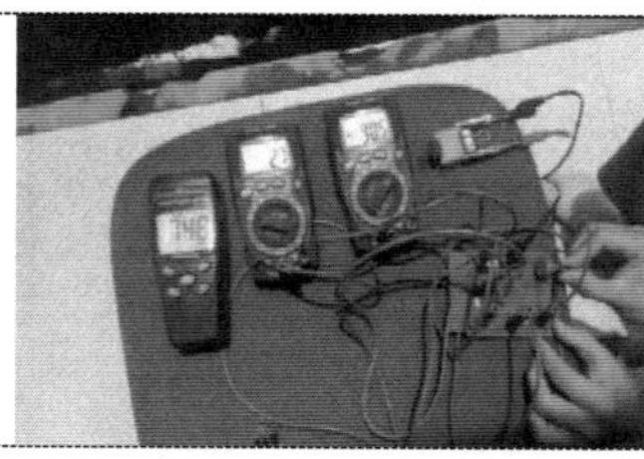

Figure 2. I-V DSSC Characteristic Test Circuit

By varying the resistance value, the current and voltage values generated from the solar cell will be obtained. In this research, the solar cell system was produced from the dye electrons which were in excited state because they were visible to light, and was injected into the conduction band of the TiO_2 semiconductor. The injection of dye electrons into the TiO_2 conduction band was facilitated by the interaction or bonding of the dye with TiO_2. If there is no interaction between the dyestuff and TiO_2, it will be more difficult for electrons to be injected. Accordingly, there will be no electron flow that could produce a solar cell system. This process could show how the effect of the dye electrons on the efficiency of the solar cell made.

3.3. *Current and Voltage Characteristics of DSSC with I-V meter*

In the I-V DSSC measurement, the working electrode and the counter electrode were arranged like a sandwich. Between the two, an electrolyte was inserted in the space between the working electrode and the counter electrode. Data collection for each sample Kappa, Lambda and Iota were conducted differently, including the day and time. The sunlight intensity for the Kappa 1:2 sample was 921 W/m^2, the Lamda 1:1 was 746 W/m^2, the Lamda 1:2 was 734 W/m^2, the Iota 1:1 was 104.7 W/m^2, Iota 1:2 was 110.3 W/m^2 and Iota 1:4 was 114.5 W/m^2.

From the results of research conducted using PEDOT Carrageenan Kappa 1:2, Lambda 1:1, Lambda 1:2, Iota 1:1, Iota 1:2, and Iota 1:4, the use of PEDOT Carrageenan Kappa 1:2 electrolyte has better results and success rates in the manufacture of DSSC than electrolytes with other types of PEDOT Carrageenan. The data of current and voltage were presented in I-V curves to determine the performance of solar cells. The voltage in the open-voltage circuit in the DSSC sample with PEDOT Carrageenan Kappa 1:2 electrolyte reached 91.8 mV, the short-circuit current (Isc) obtained reached 57.6 uA, Vmax reached 34.5 mV and Imax reached 39.2 uA, so the efficiency value obtained was 0.000333728%. According to the data, the relationship between current and voltage produces an I-V characteristic curve, with some parameters, including the short-circuit current (Isc), the open-voltage circuit (Voc), the voltage with the maximum power value (Vmax), and current that gives the maximum power value (Imax). Although this result was still very low compared to previous studies [13] which produced a Voc value of 0.439 V on the use of a mixture of PEDOT Carrageenan as an electrolyte. The current value resulted in this study was also still in the micro-Ampere order. Overall, the performance of solar cells was not good enough, especially the value of the conversion efficiency which was still very small (below 1%) [13]. The very low efficiency value was due to the very small current and voltage values. The low efficiency values obtained were associated with a very rigid PEDOT configuration in the electrolyte system. Due to the rigidity of the PEDOT chain, electron transport was not as efficient as compared to liquid systems.

The DSSC sample was made using PEDOT-Carrageenan electrolyte as a light sensitive agent at the thickness of TiO_2 paste. The preparation process included the manufacture of TiO_2 paste, synthetic dye solution from Ruthenizer, and the counter electrode preparation of PEDOT Carrageenan as electrolyte. PEDOT Carrageenan in the form of powder was dissolved in distilled water at a temperature of 70°C. After each DSSC component was ready, the DSSC assembly was carried out. The first assembly process was depositing the TiO_2 paste onto the working electrode using the doctor blade technique. Next, the working electrode was soaked with dye for 1 hour. After that, the platinum was deposited onto the counter electrode through a heating technique in a furnace for 45 minutes at a temperature of 450°C. Both electrodes were clamped with paper clips. The last step was to drip PEDOT-Carrageenan Lamda, Kappa, Iota as electrolyte about 2 drops into the space between the two electrodes. The DSSC prototype was ready to be tested.

The PEDOT-Carrageenan electrolyte in DSSC functions as a reaction generator which replaces the function of the iodolyte electrolyte which is often used in DSSC consisting of pairs of iodine and triodine as a redox in the solvent. The characteristics of the PEDOT-Carrageenan Electrolyte used in DSSC has the redox potential that is in accordance with the redox potential of the dye for maximum cell voltage and has high solubility of the solvent to support a high concentration of charge on the electrolyte. The solvent has a high diffusion coefficient for the efficient solution transportation. From the process of making DSSC, the reaction that occurred in the DSSC was the conversion of sunlight energy into electrical energy. Results indicate that the DSSC has succeeded in converting sunlight into electrical energy which was indicated by the value of current and voltage.

The developed DSSC can also be used as learning media. It can help teachers to visualize the abstract concepts to students, which will make the concepts more concrete and easier to comprehend. The components are affordable and the making steps are easy to follow thus it can be applied as hands-on activity in schools. The use of the DSSC can help students, who are not familiar with the field of material physics, to understand how to assemble a simple prototype and the processes that occur in the DSSC to produce an electric current. It can be integrated in physics concepts related to the photovoltaic concept, such as dynamic electric material. In addition, the use of DSSC as learning media can contribute to introducing and increasing students' awareness regarding the renewable energy resources [5].

4. Conclusion

The PEDOT-Carrageenan that can be used as an electrolyte system in the manufacture of DSSC is PEDOT-carrageenan Kappa 1:2. The efficiency value obtained based on the measurement of the I-V characteristics was 0.000333728%. It shows that the success rate of making DSSC with PEDOT-carrageenan electrolyte was still very low. It was owing to the equipment, materials, and work procedures used that have not met the standards. As with the use of the coarse TiO_2 powder, the thickness of the TiO2 paste layer on the working electrode and the platinum layer on the counter electrode were not uniform in each sample. Nevertheless, the developed DSSC solar cells can be used as learning media.

Acknowledgement

This publication was partially funded by International Collaboration Research Grant from Directorate General of Higher Education, Ministry of Education, Culture, Research, and Technology of the Republic of Indonesia, with funding identity: 100/SP2H/LT/DRPM/2019 and Program Kompetisi Kampus Merdeka (PK-KM) Grant of the Education, Culture, Research, and Technology of the Ministry of the Republic of Indonesia, with funding identity: 119/E1/KM.05.03/2021..

References

[1] Nasional D E 2016 *Indonesia Energy Outlook 2016* vol 2016
[2] Saehana S, Darsikin, Muslimin and Nasar 2018 Small Scale DSSC Panels Design and Its Performance *J. Phys. Conf. Ser.* **1093**
[3] Richhariya G, Kumar A, Tekasakul P and Gupta B 2017 Natural dyes for dye sensitized solar cell: A review *Renew. Sustain. Energy Rev.* **69** 705–18
[4] Saehana S, Muslimin and Abdullah M 2014 Electrochemical impedance spectroscopy study of TiO2 based solar cells *J. Renew. Sustain. Energy* **6**
[5] Grätzel M 2003 Dye-sensitized solar cells *J. Photochem. Photobiol. C Photochem. Rev.* **4** 145–53
[6] Syafinar R, Gomesh N, Irwanto M, Fareq M and Irwan Y M 2015 Cocktail Dyes from Blueberry and Dragon Fruit in The Application for DSSC *ARPN J. Eng. Appl. Sci.* **10** 6348–53
[7] Kusumandari 2009 *Meningkatkan Efisiensi Sel Surya Organik (SSO) Berbahan Klorofil-A (Bahan Alam)* (Jawa Tengah: Universitas Sebelas Maret)
[8] Maddu A, Zuhri M and . I 2007 Penggunaan Ekstrak Antosianin Kol Merah Sebagai Fotosensitizer pada Sel Surya TiO2 Nanokristal Tersensitisasi Dye *MAKARA Technol. Ser.* **11** 78–84
[9] R F K W . A H M 2010 *Kajian Pengaruh Berbagai Macam Pelarut Terhadap Karakteristik Warna Indikator Asam Basa dari Limbah Serbuk Gergaji Kayu Nangka (Artokarpus Heterophyllus L)* (Universitas Negeri Yogyakarta)
[10] Halme J 2002 Dye-Sensitized Nanostructured and Organic Photovoltaic Cells: Technical Review and Preliminary Tests *Sol. Cells*
[11] Dahlan D, Leng T S and Aziz H 2016 Dye Sensitized Solar Cells (DSSC) dengan Sensitiser Dye Alami Daun Pandan, Akar Kunyit dan Biji Beras Merah (Black Rice) *J. Ilmu Fis. | Univ. Andalas* **8** 1–8
[12] Adhitya E A, Ramelan A H and S S 2016 Sintesa Titanium dioxide (TiO2) untuk Dye-Sensitized Solar Cell dengan Antosianin Bunga Rosella (Hibiscus sabdariffa) *Indones. J. Appl. Phys.* **3** 181
[13] Ng C A and Camacho D H 2015 Polymer electrolyte system based on carrageenan-poly(3,4-ethylenedioxythiophene) (PEDOT) composite for dye sensitized solar cell *IOP Conf. Ser. Mater. Sci. Eng.* **79**
[14] Arsyad Azhar 2007 *Media Pembelajaran* (Jakarta: PT. Raja Grafindo Persada)
[15] Dimyati and Mudjiono 2009 *Belajar dan Pembelajaran* (Jakarta: Rineka Cipta)

Development of Simple Teaching Aid on Resonance Topic Using Audacity Software as Learning Media

Nurmiati[1], N Safira[1], S Saehana[1], and U Wahyono[1]*

[1]Sub-Department of Physics Education, Faculty of Teacher Training and Education Universitas Tadulako, Jl. Soekarno Hatta KM. 9 Kampus Bumi Tadulako Tondo Palu-Central Sulawesi, Indonesia

*Corresponding author: nurul.safira.7169@gmail.com

Abstract. This study aims to develop a simple prop of resonance topic using audacity software. The research was conducted based on the model from Borg and Gall, which involved only five main steps: analyzing the developed product, developing the initial product, expert validation and revision, small-scale field trials and revisions, and large-scale trials and final products. The small-scale and large-scale feasibility tests were given to 29 students of SMAN 5 Palu Class XI IPA6. The product quality data obtained were analyzed using a descriptive analysis from quantitative data into qualitative, then the average quality score was processed using category criteria. In the analysis results, the overall average score of media experts is 2.97, categorized as "Good". The overall average score from material experts was 3.33, categorized as "Very Good". While the average analysis result of the overall small-scale trials was 3.35, categorized as "Very Good", and the result of the overall large-scale trials was 3.39, categorized as "Very Good". The overall average score obtained in the transformation in table 2 shows that the simple teaching aid on resonance topic using the Audacity software developed by the researcher was feasible to be used as a medium for learning physics in the classroom.

1. Introduction

Physics is the most fundamental science because it deals with the behaviour and structure of objects [1]. In addition, Physics is a challenging and less interesting subject [2]. Many students have difficulty in learning and understanding physics concepts [3]. The process of teaching and learning physics is often confronted with abstract material [4]. Moreover, in learning activities, students are generally given the lecture method of learning [5]. With this method, students only get spoken and written words called the Verbal Method, which makes students not interested or feel compelled to study physics [6].

One suitable method for teaching physics is through demonstration [7]. It can also provide direct experience in learning [8]. The demonstration method is a teaching method by demonstrating items, events, rules, and sequences of activities, either directly or through teaching media relevant to the subject or material being presented [9,10]. The use of media (to demonstrate) is expected to play a critical role in enhancing academic performance [11].

One of the media that can be used in physics learning is teaching aid. Teaching aids are valuable instructional tools that can help make effective and interesting learning [12,13].

According to the observations made by researchers at one of the schools in Palu, namely SMAN 5 Palu, the learning process in class about resonance events was taught without any evidence to understand the resonance topic. So the researchers developed a simple teaching aid for resonance topics using Audacity software as a learning medium in SMAN 5 Palu.

Previously, a similar study was conducted by Haisy [14], who developed a resonance and doppler effect based on a PC/laptop sound card to increase high school students' motivation to learn physics. The resonance tube used was a thick, U-shaped pyrex glass tube equipped with a piston. The sound source was speakers, and Audacity software was used to show changes in amplitude when resonance occurred. Meanwhile, this study was made simpler, in which the Audacity software-assisted resonance teaching aid was used as a learning medium. The resonance tube used was a 90 cm long scrap iron pipe equipped with a scale piston. Researchers used speakers, but the sound was controlled with a frequency generator application on an Android cellphone.

The benefit of this research in the education field is that it can produce learning media. Learning media in the form of teaching aids can help teachers make the teaching and learning process more effective and efficient in the classroom.

2. Method

2.1. Research Design

The type of research used in this research was Research and Development (R&D). This research was conducted based on the developmental research model from Borg and Gall, involving only five main steps: analyzing the to-be-developed product, developing the initial product, expert validation and revision, small-scale field trials and revisions, and large-scale trials and final product. Tools and materials used for this research were provided in Table 1, while the design was presented in Figure 1.

Table 1. Tools and materials

Tools	Explanation and Function
Scrap iron pipe	This 90 cm long iron pipe was found in the wreck and functions as a resonance tube.
Used board	The used board here was the leftover board from a carpenter that was no longer used, and its function was to make a resonant tube holder and a speaker holder.
Used sandals	Its function was to make rubber from the piston.
Piston	Its function was to adjust the length of the air column.
Speaker	It served as a sound/sound source.
An android phone equipped with the Frequency Generator application	The frequency Generator only produces tones between 1 Hz-22000 Hz, and this application was used as a sound source controller.
Laptop with Audacity Software	It was used to analyze sound frequency.

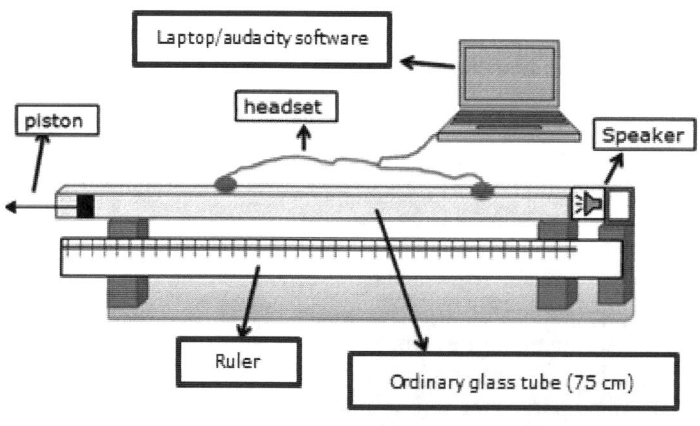

Figure 1. Resonance props scheme

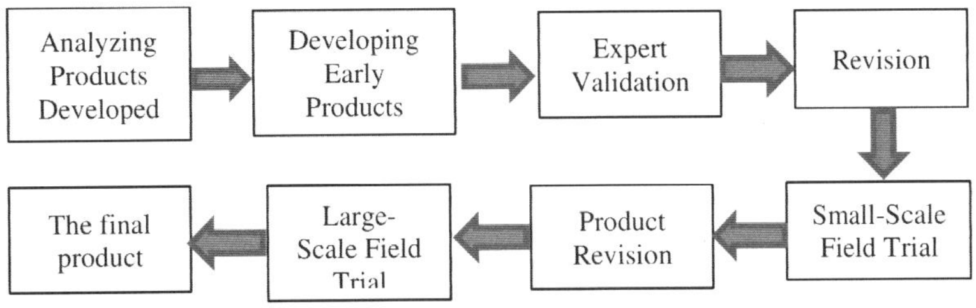

Figure 2. Research flowchart

This research was conducted in SMAN 5 Palu at Class XI IPA6, amounting to 29 people, five students became small-scale respondents, and 24 students became large-scale respondents. The type of data consisted of quantitative and qualitative data. Quantitative data was in the form of product assessment scores and the feasibility of simple resonance teaching aids using Audacity software based on the results of a validity test questionnaire with a Likert scale in numbers 1, 2, 3, and 4. Qualitative data was the evaluation from the validator (responses, input, suggestions and criticisms) listed in the questionnaire or directly used as a consideration in revising the simple resonance teaching aids.

Assessing the feasibility of a product can be seen from the feasibility indicators. Media expert assessors, material experts and students have their respective eligibility indicators. The indicators for the feasibility of media consist of attractiveness, placement suitability of tools and materials, ease of use, teaching aids can run well, ease of transfer, ease of storage, facilitation to teaching and learning activities, increase of student knowledge, and the durability of the teaching aids. The material expert's eligibility indicators consist of showing a clear resonance event, showing changes in amplitude when resonance occurs, measuring the resonance frequency, students easily understanding resonance events, the suitability of teaching aids in learning, and increasing students' competence. The feasibility indicators for students are attractiveness, proper shape and location of tools and materials, ease of understanding resonance events, emerging enthusiasm in learning, ease of operation, facilitation to understand resonance, the durability of props, simplicity, and ease of management.

Arikunto [15] stated that finding out the final score rating on each item of the assessment questionnaire was divided by the total score obtained was divided by the number of respondents who answered the assessment questionnaire. Then, the average score obtained was transformed in Table 2 to see the feasibility of a product.

Table 2. Media eligibility score interpretation criteria

Achievement percentage	Average score	Interpretation
76% – 100%	4	Very decent
51% – 75%	3	Decent
26% – 50%	2	Decent enough
0% – 25%	1	Less decent

2.2. Working Principle of Props

The working principle of this teaching aid is, first, open the Audacity software, then turn on the speaker (which is set by the generator frequency) and place it at the end of the first tube. The resonance tube is equipped with a piston placed at the end of the second tube, where the piston is pulled slowly after recording on audacity begins. Then, there will be a sound getting louder and louder, and the loudest sound is the first resonance event. Measuring sound frequency can be analyzed with audacity software by selecting the spectrum of the sound signal using the selection tool. Then on the menu bar, select "Analyze" and press "Spectrum Plot". After that, take the highest peak in the "Frequency Analysis" picture, the first basic tone or resonance.

Journal of Physics: Conference Series **2126** (2021) 012005 doi:10.1088/1742-6596/2126/1/012005

The illustration of the displayed results can be seen in Figure 3. Determine the wavelength, the speed of wave propagation and the value of the frequency of sound using the following formula.

f_{hp}(Generator Frequency) = 250 Hz

L = 33 cm = 0.33 m

$$\lambda = \frac{4L}{n} = \frac{4(0{,}33\ m)}{1} = 1.32\text{ m}$$

$$v = f_{hp}.\lambda = 250 \times 1.32 = 330\text{ m/s}$$

$$f = \frac{v}{\lambda} = \frac{330\ m/s}{1{,}32\ m} = 250\text{ Hz}$$

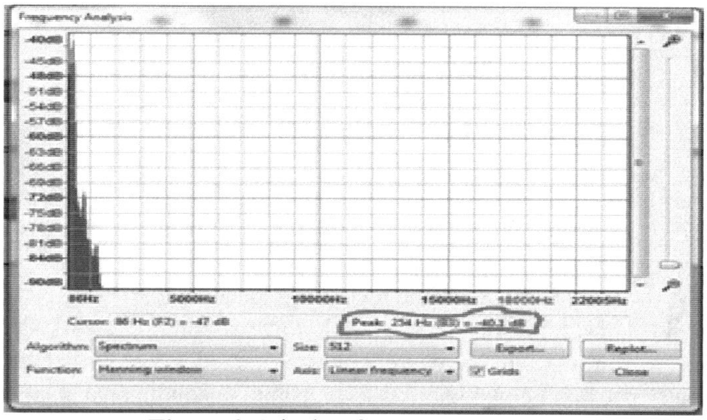

Figure 3. Display frequency analysis

3. Result and Discussion

In operating this simple resonant material prop, using constant motion is a must. Pulling or pushing the piston should be done slowly. Observers of resonance events on props must have good hearing and be careful. Therefore, the frequency value calculated by the formula is the same as the source frequency value in the Generator Frequency application.

Resonance events are commonly used to determine the speed of sound in the air. The theoretical value of the speed of sound in the air is 340 m/s. In comparison, the value obtained using a simple resonance topic was 330 m/s at a frequency of 250 Hz. It means that the value was close to or almost the same. In line with research conducted by Agustina [16], the value of the speed of sound in the air obtained was close to the theoretical value. The results of measurements using simple props on resonance topics can be seen in Table 3.

Table 3. Measurement results using a simple resonance visual aid

f_{hp} (Hz)	Number of Resonance (n=1,3,5,...N)	L (m)	f_A (Hz)	f_h (Hz)	λ (m)	v (m/s)
250	1	33	254	250	1.32	330
550	1	47	551		**1.88**	**1034**
	2	79	551	550	1.05	577.5
750	1	32	745	**750**	**1.28**	**960**
	2	65	746		0.87	**652.5**
	3	80	746		0.62	**480**

It can be seen in Fig. 11 that the relationship between frequency and the number of observed resonance events is that the greater the frequency, the more resonance events are observed. Fig. 12 also shows that the relationship between frequency and the length of the air column is, the greater the frequency, the greater the length of the air column. Fig. 13 shows the difference between generator frequency, audacity frequency, and count frequency.

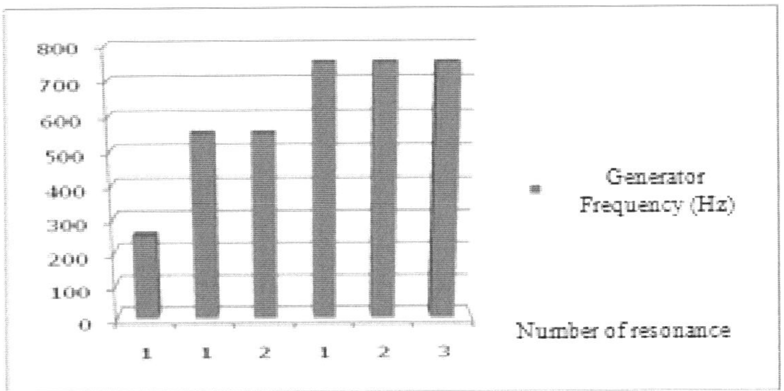

Figure 4. Frequency relationship diagram with the number of observed resonances

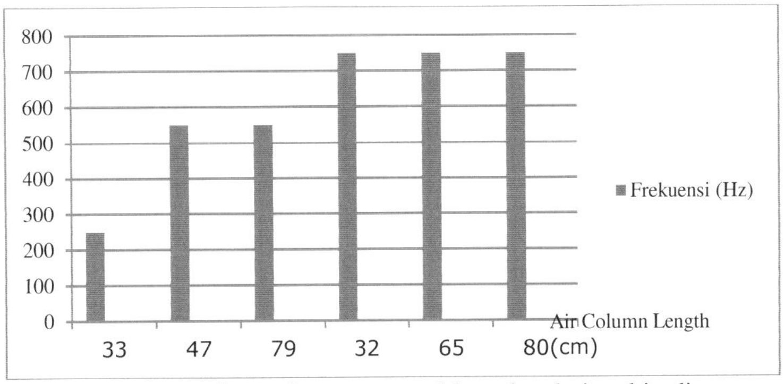

Figure 5. Air column frequency and length relationship diagram

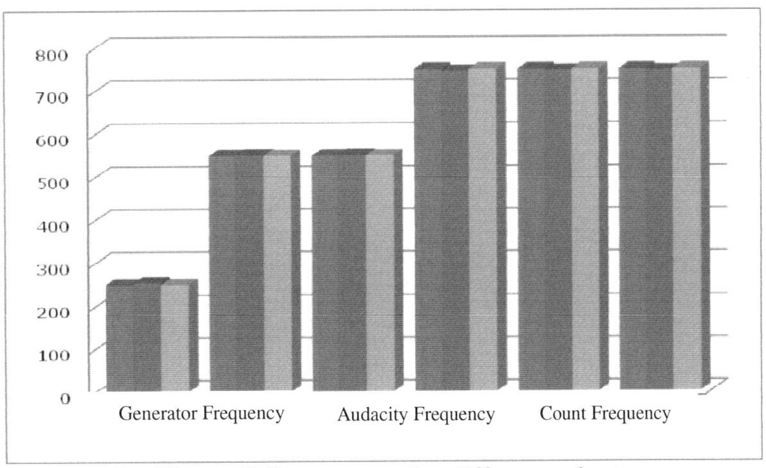

Figure 6. Frequency value difference chart

The analysis results of media experts obtained an average score of 2.97 for all aspects and were categorized as "Good" as in Table 4. For the material expert analysis results, the average score for all aspects was 3.33 and categorized as "Very Good". From the average score obtained, which was then transformed into table 5, the achievement percentage was around 51%-75% with a "decent" interpretation. Hence, it can be stated that this simple resonance topic is very good and deserves to be tested in the field with revisions.

Table 4. Validation results by media experts

No	Aspect	Average Score	Category
1	Props design	3.33	Very Good
2	Quality and effectiveness of props	3.50	Very Good
3	Operation and performance of props	3.00	Good
4	Clarity of usage instructions	2.00	Less
5	The ability of the media to increase knowledge	3.00	Good
6	Props resistance	3.00	Good
	Overall Average	2.97	Good

Table 5. Validation results by material experts

No	Aspect	Average Score	Category
1	The relationship between props and resonance topic	3.50	Very Good
2	Educational value	3.50	Very Good
3	Physics content	3.00	Good
	Overall Average	3.33	Very Good

After doing the validation, the props that have been developed were tested on students. The results of the small-scale trial analysis obtained an overall average score of 3.35 in the "Very Good" category as in Table 6. Likewise, the analysis results on large-scale trials obtained an overall average score of 3.39 with the "Very Good" category as in Table 7. From the average score, which was then transformed into Table 2, the achievement percentage was around 51%-75%, with a proper interpretation. Therefore, it can be said that the simple resonance teaching aid that was developed is very good and suitable to be used as a medium for learning physics during the teaching and learning process in the classroom.

Table 6 . The results of the analysis of the average small-scale student responses

No	Aspect	Average Score	Category
1	The physical appearance of props	3.5	Very Good
2	Motivation to learn by using props	3.4	Very Good
3	Aspects of the operation of the props	3.2	Good
4	Props quality	3.3	Very Good
	Overall Average	3.35	Very Good

Table 7. The results of the analysis of the average large-scale student responses

No	Aspect	Average Score	Category
1	The physical appearance of props	3.34	Very Good
2	Motivation to learn by using props	3.40	Very Good
3	Aspects of the operation of the props	3.44	Very Good
4	Props quality	3.41	Very Good
	Overall Average	3.39	Very Good

The teaching aid has advantages and disadvantages in the research development process to the field trial stage for sure. The advantages of these teaching aids include that the components are easy to find because they are commonly found in the surrounding environment, and we usually use them every day. In addition, in its use, this simple resonance visual aid can record and calculate the frequency automatically and see the frequency automatically through the analysis frequency graph, thus making it easier for students to understand the resonance event. The drawback of this teaching aid is that it cannot automatically calculate the sound speed in the air. In addition, to use this simple resonance topic prop, it must be done with two or three people, and the use of the audacity software must be in a quiet room so as not to be affected by other sound sources.

4. Conclusion

The simple teaching aid of resonance topic using audacity software that has been developed is overall good and suitable to be used as a physics learning medium during the teaching and learning process based on assessments by media experts, material experts and students. This simple resonance teaching aid makes it easier for students to understand the resonance topic taught by demonstrating it in front of the class.

Acknowledgement

We thank kindly Hibah Penelitian Terapan Kompetitif Nasional DIKTI and Program Kompetisi Kampus Merdeka (PK-KM) of Ministry of Education, Culture, Research and Technology of The Republic of Indonesia, under the funding 302/E4.1/AK.04/PT/2021 and 119/E1/KM.05.03/2021, respectively.

References

[1] Yudhittiara R F, Hindharto N and Mosik M 2017 *UPEJ Unnes Phys. Educ. J.* **6** 81–89
[2] Alatas F, Mulhayatiah D and Jahrudin A 2015 *J. Penelit. dan Pembelajaran IPA* **1** 60–75
[3] Ekici E 2016 *J. Educ. Pract.* **7** 95–107
[4] Mufit F, Asrizal A and Puspitasari R 2020 *J. Penelit. Pengemb. Pendidik. Fis.* **6** 267–78
[5] Nesi M and Akobiarek M 2018 *BIOEDUSAINS J. Pendidik. Biol. Dan Sains* **1** 80–94
[6] Charli L, Ariani T and Asmara L 2019 *SPEJ (Science Physic Educ. J.* **2** 52–60
[7] Saehana S, Ali M and Supriyatman S 2019 *J. Pendidik. IPA Indones.* **8** 241–246
[8] Nahdi D S, Yonanda D A and Agustin N F 2018 *J. Cakrawala Pendas* **4**
[9] Ramadhan N and Surya E 2017 *Int. J. Sci. Basic Appl. Res. (IJSBAR* **34** 62–68
[10] Rohman A, Komang Werdhiana I and Saehana S 2021 *J. Phys. Conf. Ser.* **1760**
[11] Benson A and Odera F 2013 *Int. J. Inf. Commun. Technol. Res.* **3**
[12] Haryadi R, Vita M, Utami I S, Ihsanudin I, Setiani Y and Suherman A 2019 *J. Phys. Conf. Ser.* **1157** 32006
[13] Izzah A and Saehana S 2021 *J. Phys. Conf. Ser.* **1760**
[14] Haisy M C, Astra I M and Handoko E 2015 *In prosiding seminar nasional fisika (e-journal), SNF2015-II, 4*
[15] Arikunto S 2013 *Prosedur Penelitian Suatu Pendekatan Praktik* (Jakarta: Rineka Cipta)
[16] Astuti I A D 2016 *UPEJ Unnes Phys. Educ. J.* **5** 18–24

Journal of Physics: Conference Series **2126** (2021) 012006 doi:10.1088/1742-6596/2126/1/012006

Refractor telescope design using web camera

M A Karaeng[1], S Saehana[1], and U Wahyono [1*]

[1]Sub-Department of Physics Education, Faculty of Teacher Training and Education Universitas Tadulako, Jl. Soekarno Hatta KM. 9 Kampus Bumi Tadulako Tondo Palu-Central Sulawesi, Indonesia.

*Corresponding author: uwahyono@gmail.com

Abstract. The purpose of this study was to develop a refractor telescope using a web camera as learning media. It was the Research and Development (R&D) study which referred to the model modified by Sugiyono. The product was evaluated by a media expert and undergraduate students of Sub-Department of Physics Education, Tadulako University using questionnaires with a 4-point Likert scale. The product quality data obtained were analyzed using descriptive analysis. The quality of the product based on the analysis of the media expert's assessment was good with the average score of 2.92. Meanwhile, the results of student responses to the limited trial obtained a score of 3.20 which indicated the agreement of usability. The test results and questionnaire analysis showed that the refractor telescope using the developed Web Camera (WebCam) is feasible to be used as learning media.

1. Introduction

Physics requires a complex of scientific skills. Physics learning requires students to observe, measure, and analyse natural phenomena, scrutinize them with their knowledge, and draw conclusions based on these processes [1]. It involves the skills in the three domains of learning, namely cognitive, affective and psychomotor [2,3]. Students need sufficient cognitive skill to understand the theories, principles and concepts of physics. Meanwhile, the affective and psychomotor skills are required especially during hands-on activities [4,5]. These skills enable students to work in active and collaborative learning during experiments [6,7]. However, the science learning activities mostly still prioritize the achievement of cognitive domain [8,9].

Accordingly, science learning cannot be separated from laboratory activities since it was the media for the development of scientific skills [10–12]. Students need practical work to solve problems, especially those that are closely related to everyday life. In addition, there are some abstract physics concepts that are difficult to understand. Experiments help students to not only remember the concepts, but also understand its application in real life. Direct observation and experiment will give students the opportunity to see and prove the theory that they have learned. This will certainly increase creativity, and skills and consequently boost their learning motivation [13,14].

The use of pictures in learning is imperative, both in the realm of understanding new words or new concepts [15–18]. The use of pictures and images in science becomes more crucial owing to many natural phenomena, objects or processes are too small, too far, too fast and too slow to observe and study directly [19]. In addition, since the scientific process consists not only of understanding and discovery, but also of communication [19], the use of any type of visual media allows the visualization and illustration of abstract concepts [20–22].

On that account, this study aimed to develop a refractor telescope using a web camera as learning media. The relevant research has been conducted by Robin et al [23] who designed a telescope with two webcams to acquire data. Our study only used one webcam thus it would be more affordable. The product is also easy to move and light thus we hope that it can be used as a tool in science learning.

2. Methods

2.1. Types of Research

This research belongs to the Research and Development (R&D), which is used to produce certain products and to test the effectiveness of these products [24].

Refractor telescope design research using a web camera for learning media adapted the steps of the development research model by Sugiyono [24]. The general research steps are shown in Figure 1.

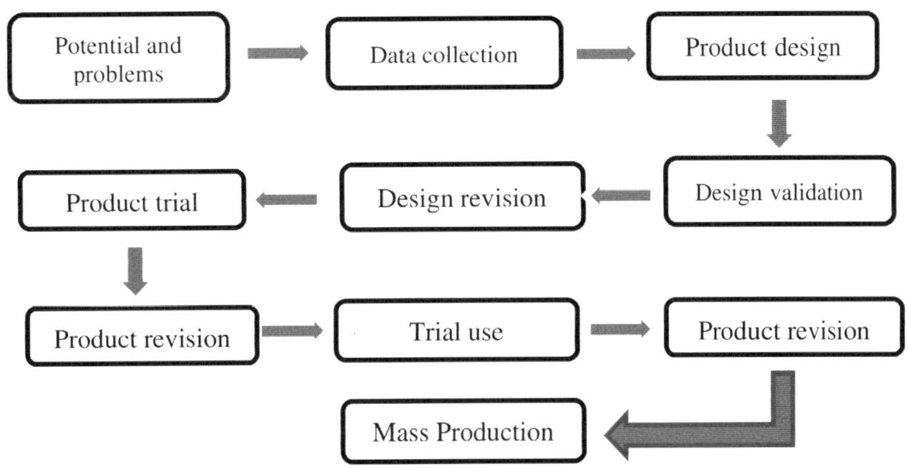

Figure 1. Steps of research and development method

2.2. The steps in the research are only up to the product revision stage

The preliminary activity was to investigate the potential opportunities and problems. The next step was gathering the materials needed to make media. The third step was product manufacturing. We started with drawing the circuit of the device, and eventually developed the product. After that, the developed product was sent to the validator for assessment. This validation was used to find out whether the product made is worth testing or not. Validation was done through discussion with the experts. The expert here was a media expert who is competent in their fields. Revision was conducted to the product based on the evaluation before the product underwent field-test. The revision step was the last step of this research.

2.3. Field-test Data Collection Procedures

Since the field-test was only a limited testing, the subjects of this study involved only 11 undergraduate students of the Sub-Department of Physics Education in Tadulako University who enrolled in 2015 academic year. This evaluation aimed to assess the feasibility of the practicum tool using a questionnaire [25].

The analytical technique used to analyze the data from the validation results is the calculation of the average value. It was based on Arikunto's opinion that the final score for each research questionnaire item was the average score [26], then use the scores to allocate the product into categories based on the interpretation of Widoyoko [27].

3. Results and Discussion

3.1. Tool Quality Assessment by Expert

The aspects assessed by the expert included the quality of the tools and the effectiveness of the tools, the availability of second-hand goods that are easily affordable, and the suitability of the tools with the concept of physics. The aspects assessed include the aspects of tool quality and tool efficiency, the affordability of tools and materials as well as the suitability of the tool with physics concepts. The results of the assessment of all aspects were in the good category. The results of the assessment by experts are presented in a diagram as shown in Figure 2.

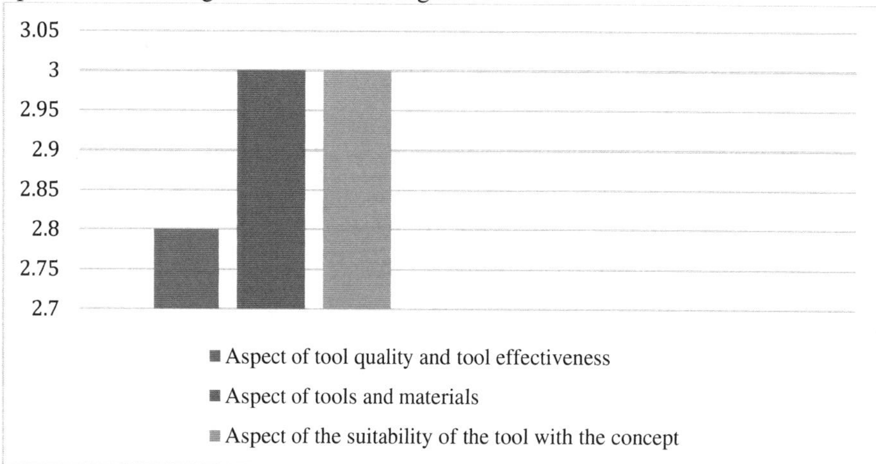

Figure 2. Scores of media expert' assessment

From the assessment of media experts, there are several notes that were still considered lacking, including the criteria for the ease of tool operation in accordance with the working principles, the ease in the tool management, the portrayal of abstract to more concrete abstract by the tool, and the conceptual knowledge demonstrated by the tool. The expert pointed out that the tool was still sensitive during its application. The users should be careful when adjusting the focus of the telescope and webcam. Moreover, there were several inputs stating that the tool was not yet clear in showing abstract concepts to be more concrete. In addition, the expert suggested placing the camera on the lens permanently so that it is easy to operate.

3.2. Response Results on Limited Test

Limited test was conducted to measure some aspects, including the physical appearance of the practicum tool, aspects of learning motivation with tools, aspects of tool operation, aspects of tool quality and aspects of observation. The assessment on the limited trial was carried out after the product underwent the revisions based on the expert validation. The results of the responses in the limited trial showed that students were interested in the developed telescope. They were enthusiastic because they have never used a tool similar to the product. The data indicated that most agree to use the refractor telescope using a web camera designed by the researcher in their learning activities. The results of the students' responses are displayed in Figure 3. Generally, students suggested making the tool more interesting in appearance and usability.

The general suggestions and criticisms from the responses included:

1) Tool appearance should be made more attractive and portable
2) The tool should be further developed so that it can be used in bright places or rooms.

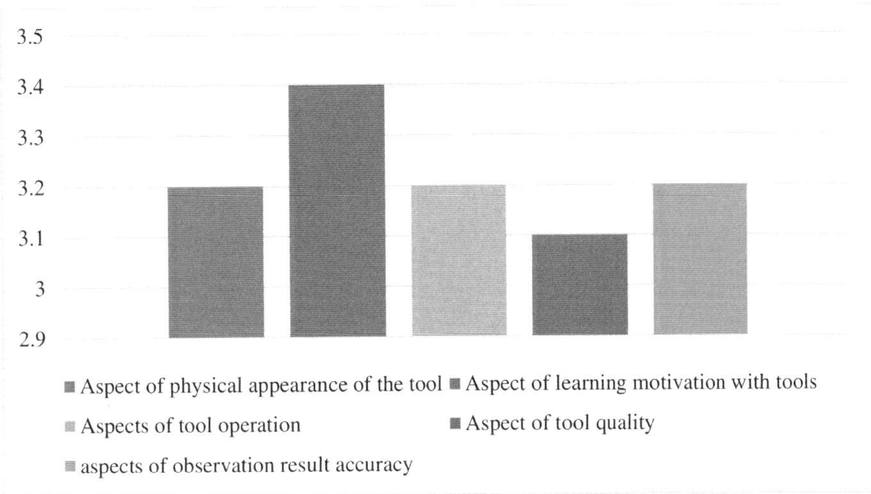

Figure 3. Assessment diagram by physics education students

3.3. The Developed Refractor Telescope

The refractor telescope designed using a web camera (webcam) was presented in Figure 4 while the working principle scheme was displayed in Figure 5.

Figure 4. Refractor Telescope Using a Web Camera (webcam).

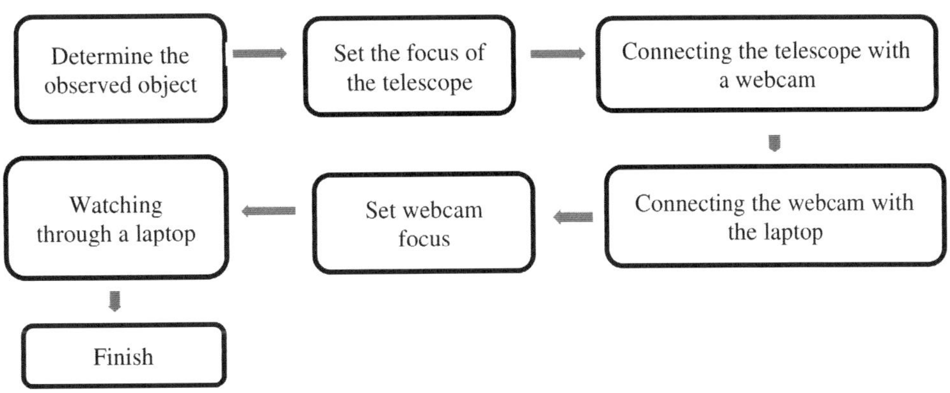

Figure 5. Tool working principles

Observations of refractor telescopes using a web camera (webcam) can be seen in Figure 6.

Figure 6. Observations of the moon with refractor telescopes using webcams

Since the used of digital images can help students to overcome misconception [19], this tool can be used as a substitute for telescopes in laboratories that are no longer functioning. The students also could follow the stages and working principles of this design and make their own telescope.

4. Conclusion

Based on data analysis and discussion, a refractor telescope has been made using a web camera (webcam). The results of the assessment given by experts and physics education students on the feasibility of the product indicated that the tool has met the criteria of the feasibility of the product to be used as learning media.

Acknowledgment

We highly acknowledge the support of the Program Kompetisi Kampus Merdeka (PK-KM) Grant of the Education, Culture, Research, and Technology of the Ministry of the Republic of Indonesia, with funding identity: 119/E1/KM.05.03/2021.

References

[1] Weisberg M 2012 *Simulation and Similarity: Using Models to Understand the World* (Oxford, UK: Oxford University Press)

[2] Chang C Y, Lai C L and Hwang G J 2018 Trends and research issues of mobile learning studies in nursing education: A review of academic publications from 1971 to 2016 *Comput. Educ.* **116**

[3] Sutrisno V L P and Siswanto B T 2016 Faktor-Faktor Yang Mempengaruhi Hasil Belajar Siswa Pada Pembelajaran Praktik Kelistrikan Otomotif SMK di Kota Yogyakarta *J. Pendidik. Vokasi* **6**

[4] Plummer L, Smith L, Cornforth E and Gore S 2021 Teaching psychomotor skills in a virtual environment: An educational case study *Educ. Sci.* **11**

[5] Zach S and Rosenblum H 2021 The affective domain-A program to foster social-emotional orientation in novice physical education teachers *Int. J. Environ. Res. Public Health* **18**

[6] Vidergar A B 2013 *Fictions of Destruction: Post-1945 Narrative and Disaster in the ollective Imaginary* (Stanford University)

[7] Jette D U, Macauley K and Levangie P K 2020 A Theoretical Framework and Process for Implementing a Spiral Integrated Curriculum in a Physical Therapist Education Program *J. Phys. Ther. Educ.* **34**

[8] Malik A, Khasanah U, Agustina R D and Zakwandi R 2019 Which One Is Better Hands on or Phet Simulation on Harmonic Oscillation *Edusains* **11** 264–78

[9] Creswell R 2014 *Educational research: Planning, conducting and evaluating quantitative and qualitative research* (Sydney, Australia: Pearson)

[10] Wiwin E and Kustijono R 2018 The use of physics practicum to train science process skills and its effect on scientific attitude of vocational high school students *Journal of Physics: Conference Series* vol 997

[11] Abd-El-Khalick F, Boujaoude S, Duschl R, Lederman N G, Mamlok-Naaman R, Hofstein A, Niaz M, Treagust D and Tuan H L 2004 Inquiry in science education: International perspectives *Science Education* vol 88

[12] Sheeba M N 2013 An Anatomy of Science Process Skills In The Light Of The Challenges to Realize Science Instruction Leading To Global Excellence in Education *Educ. Confab* **2**

[13] Chen J C 2020 Developing a cycle-mode POED model and using scientific inquiry for a practice activity to improve students' learning motivation, learning performance, and hands-on ability *Interact. Learn. Environ.*

[14] Wen-jin K, Chia-ju L and Shi-an L 2012 Promoting Female Students' Learning Motivation towards Science by Exercising Hands-On Activities. *Online Submiss.*

[15] Paji K K, Talepasand S and Kianersi F 2018 The Role Of Pictures And Gestures As Nonverbal Aids In Learning English Words As The Second Language In Preschool Children *Int. J. Manag. Appl. Sci.* **4** 47–50

[16] Rowe M L, Silverman R D and Mullan B E 2013 The role of pictures and gestures as nonverbal aids in preschoolers' word learning in a novel language *Contemp. Educ. Psychol.* **38**

[17] Kokab K P, Siavash T and Farahnaz K 2019 The role of gestures and pictures as nonverbal teachning/learning aids in affecting female and male preschoolers' learning of new words *Stud. Learn. Instr.* **11** 128–45

[18] Triacca S 2017 Teaching and Learning with Pictures the Use of Photography in Primary Schools *Proceedings* **1**

[19] Bell R L, Park J C and Toti D 2004 Digital images in the science classroom *Learn. Lead. With Technol.* **31**

[20] Ismail A, Gumilar S, Amalia I F, Bhakti D D and Nugraha I 2019 Physics learning media based Augmented Reality (AR) for electricity concepts *Journal of Physics: Conference Series* vol 1402 (IOP Publishing Ltd)

[21] Suprapto N, Nandyansah W and Mubarok H 2020 An evaluation of the "PicsAR" research project: An augmented reality in physics learning *Int. J. Emerg. Technol. Learn.* **15** 113–25

[22] Suprapto N and Nandyansah W 2021 PicsAR: A Physics Visualisation to Enhance Students' Thinking Skills in Abstract Concepts *Journal of Physics: Conference Series* vol 1805 (IOP Publishing Ltd)

[23] Robin G, Shukla A K and Jolivot R 2019 Remotely-controlled low-cost digital telescope *Eng. J.* **23** 141–56

[24] Sugiyono 2012 *Metode Penelitian Pendidikan: Pendekatan Kuantitatif, Kualitatif, dan R&D.* (Bandung: Alfabeta)

[25] Endang 2011 *Riset Terapan Bidang Pendidikan dan Teknik* (Yogyakarta: UNY Press)

[26] Arikunto S 2011 *Prosedur Penelitian Pendekatan Suatu Praktik* (Jakarta, Indonesia: Rineka Cipta)

[27] Widoyoko 2012 *Teknik Penyusunan Instrumen Penelitian* (Yogyakarta: Pustaka Pelajaran)

Journal of Physics: Conference Series **2126** (2021) 012007 doi:10.1088/1742-6596/2126/1/012007

Cultivating science process skills among physics students using PhET simulation in teaching

S Ruwiyah[1*], N F A Rahman[1], A R A.Rahim[2], M Y Yusof[2], S H Umar[2]

[1]School of Education, Faculty of Social Sciences and Humanities, Universiti Teknologi Malaysia, 81310 Skudai Johor.
[2]Johor Matriculation College, Jalan Payamas 84900 Tangkak Johor.

[*]Corresponding author: samijah_ruwiyah@yahoo.com

Abstract. This paper concerns with the use of simulation as a pedagogical tool for a physics teaching strategy. This action research was conducted to investigate the scientific skills as a result of modelling learnt using PhET simulation. The cycle of action research involved in this study is planning, acting, observing and reflection. Through combining physics simulation with Science Process Skills (SPS) and Manipulative Skills (MS) to deliver real experimenting experience, the discussion is further to make connection with students' achievement in Physics. One physics class involved in this study consist of 36 matriculation college students. Prior the study, students were briefed on the procedures for conducting the Harmonic Motion experiment using the PhET Simulations. Following this, students must completed the practical report for analysis, data interpretation and discussion. At the end of the teaching, a set of formative test was used to evaluate students' mastery of Physics concepts after conducting the Simple Harmonic Motion simulation experiment. The results showed an increase in the percentage of frequency of study participants practicing Science Process Skills and Manipulative Skills (Conducting Experiments). In the Formative Test, the study participants showed a high average score of 13.9 marks, which is 93%. This study is able to develop the advantage of multimedia for physics laboratory works and explore its potential to strengthen scientific skills.

1. Introduction

Education in Malaysia as enshrined in the National Education Philosophy is a continuous effort to further develop the potential of individuals in a comprehensive and integrated manner to produce a balanced human being, in terms of intellectual, spiritual, emotional, and physical. The Primary School Science Curriculum Standard (KSSR) and the Secondary School Science Curriculum (KSSM) were developed to produce the desired human beings [1], in the National Philosophy of Education. KSSR and KSSM introduce six pillars that focus on the formation of balanced human capital in terms of physical, emotional, spiritual, intellectual, and social. One of the six pillars is the pillar of science which emphasizes the mastery of scientific knowledge, skills, and attitudes.

The Matriculation Program emphasizes inquiry as a way for enhancing students' ways of knowing in subject matter discipline. In the process of inquiry, problem-solving scientific skills and thinking skills need to be used. Scientific skills are cultivated in any activity by performing the desire scientific method. Scientific skills consist of science process skills and manipulative skills [2,3].

Science process skills such as making accurate observations, taking measurements in the right way, using appropriate measuring tools, drawing conclusions, making hypotheses, predicting, interpreting data, and controlling for variables are further developed during practical sessions. Such processes are integrated to help students improve Higher Order Thinking Skills such as analytical, critical, and creative thinking skills. These skills are very important to prepare students for the challenges ahead in the 21st century [4].

Manipulative skills such as operating instruments, setting up apparatus properly and drawing diagrams [5] can be extended through practical sessions. Laboratory experiments are designed to encourage students to have an inquisitive mind. It requires students to actively participate in science process skills before, during and after experiments by preparing pre-reports, making observations, analysing results, and drawing conclusions [4]. A combination of science process skills and manipulative skills can enhance students' high -level thinking skills.

The spread of the Covid-19 pandemic in Malaysia has seen a very drastic shift in new norms, including in the aspect of physics education. At the ministry level, teachers are require to adopt the hybrid Teaching and Learning (T&L) as the counter measure due to pandemic [6]. There are three modes of teaching and learning in the science stream at matriculation, namely lectures, tutoring and practical. This study focuses on the changes made in designing the practical lessons. Practical Physics subject often implemented through face-to-face because it allows students to master the necessary skills and ways of thinking as scientist [7]. Unfortunately, with Covid-19 pandemic situation, face-to-face teaching and learning cannot be implemented to reduce the risk of infection since June 2020.

These changes are not unique in Malaysia only, since other countries in the rest part of the world are force to make drastic changes in teaching subject that require laboratory assessment. For instance, in United Kingdom, [8] reviewed challenges among tertiary education when it comes to transform traditional to online learning. Some of the similarities with Malaysia are the acceleration of multimedia and opportunity to popularize remote laboratory practice. In this study, from the simulation method, students can still conduct experiments and practice scientific skills even without physical laboratory facilities. For instance, works done by [9] report on collaborative strategy between asynchronous and synchronous to conduct laboratory activities during pandemic. One of the advantage highlighted by [9] with remote lab is the opportunity to activate student's agency toward learning. [10] added that remote laboratory is way influential to activate practical skills of science and become useful resources that enhance accessibility to learning.

The Physics Education Technology (PhET) simulation program was developed by the University of Colorado at Boulder, USA [11]. Its' main purpose is to facilitate teachers and students to do laboratory activities in the classroom. PhET simulation is a user-friendly simulation because it can be run using an existing web browser. Simulations in PhET are available free of charge and can be accessed through the website http://phet.colorado.edu.

PhET simulations use graphics with animated visuals [12]. These simulations are designed in an animated, interactive, and game-like form where students learn through exploration. PhET simulations combined the results of research and variables performed by students so that students can relate the phenomena of daily life with the correct Physics concepts. Thus the use of PhET simulation is recognized for its validity worldwide [11].

Based on the information from the University of Colorado Boulder, PhET simulations were developed using the following design principles. Firstly, to encourage scientific research, second, to provide interaction, third, to create something invisible to the naked eye, fourth, to feature visual mental models, fifth, to feature some representations (examples, object movements, graphs, numbers, etc.), the sixth connects users with the real world, the seventh gives users on implicit guidance in exploration, and the eighth makes flexible simulations and can be usable in a variety of educational situations.

Several tools in PhET simulation also provide an interactive experience [13], such as: first click and drag to interact with simulation features, secondly, launch the cursor to increase and decrease parameters and thirdly, make measurements in experiments with various instruments. Users who interact with the tool are immediately notified on the feedback of the impressions of the changes that they have made. This allows them to investigate cause-and-effect relationships and answer scientific questions through simulated exploration.

Physics is one of the core subjects for the Matriculation Program of the Ministry of Education Malaysia (MOE). The KPM Matriculation Program is divided into two systems, the Two Semester System (SDS) and the Four Semester System (SES). The researchers focused this action research on SES Module 1 students as the researcher was also teaching and supervising the class involved.

Scoring for Physics subjects is divided into two, written assessment 60% and continuous assessment (PB) 40%. Weighting for PB, 10% of practical test, 20% of practical report and 10% of assignment. Assessment of practical tests and practical reports involves the mastery of scientific skills by students.

Below are the reflections obtained in previous lesson when teachers are forced to adopt emergency remote teaching. The temptation to focus on educational outcomes during pandemic time, with learning outcome for face-to-face learning has created dissatisfaction among teachers and students with the learning experiences. Teachers on the other hand is force to create creative problem solving to generate various possible solution to meet the need of educational standard.

Lecturers' Reflections

Following the increase in COVID-19 pandemic cases and the implementation of the Conditional Movement Control Order (MCO) by the National Security Council involving almost all states in Malaysia, the Ministry of Education (MOE) has decided that all Matriculation colleges implement a blended teaching and learning (T&L) process [6]. This blended teaching and learning practices a combination of face-to-face and non-face-to-face teaching and learning [14]. For practical mode, the time allotted is two hours per week for each class. Due to the large number of students coupled with the laboratory space and the limited number of laboratories, online face-to-face teaching and learning is implemented.

For teaching and learning practical mode, the supervising lecturer provides a video of the process of conducting the experiment up to obtaining readings from the experiments conducted. This video is uploaded through the YouTube platform to make it easier for students to download practical videos. Instructions to guide students in the teaching and learning process are uploaded through the Google Classroom (GC) platform. Finally, students will complete a practical report and uploaded it through the GC platform for their supervising lecturer review.

The researchers are aware that this teaching and learning process as a whole does not provide opportunities for students to practice scientific skills. Science process skills such as classifying, predicting, communicating, controlling variables and experimenting [4], cannot be practiced by students. Researchers feel that this needs to be improved because of this situation will have a great impact on students' ability to improve their high-level thinking skills such as having analytical, critical, and creative thinking skills which is an important agenda in nation building and to face progress and challenges and global change waves in the 21st century.

Students' Reflection

Through the Google Classroom (GC) platform, researchers have provided a special space for students to write down their weekly reflections. This reflection space is open to no compulsion or obligation to be written by the students. In addition, to encourage the writing of sincere reflection from the hearts of the students, no name or matriculation number is recorded. The information of the reflection writer remains the secret of the students.

Some scraps written by the students showed their difficulties in the teaching and learning process that they faced.

'Wasting time waiting for videos or loading work in google classroom'
'Learning using video'
'Ineffective learning.'
'I became more depressed'
'Difficult to download learning videos given by lecturers'

Looking at these reflections, thereby this study proposes as a solution to overcome issues with access to cultivate science process skills. The aims are to provide experience for students to do virtual experiments using PhET Simulation and to strengthen students' thinking skills in increasing their intellectual power.

Specific objectives:

1. To strengthen students' science process skills, through virtual inquiry.
2. To strengthen students' manipulative skills.
3. To strengthen students' knowledge and understanding of the facts and Physics concepts

2. Method

Basically, every action research will involve two elements, namely research and action. These two elements are then developed and detailed into research models that action researchers follow. This study used action research model by Kurt Lewin 1947 [4]. Kurt Lewin (1890-1947) is known as the father of action research because he is considered to be the first person to mention the term action research through an article entitled Action Research and Minority [15]. He proposes a spiral process involving steps of planning, action and fact finding [16].

2.1 Planning

In this action research, the research instruments used are scientific skills checklists, student practical report rubrics and formative tests. Scientific Skills checklist is used to obtain information from students, science process skills and manipulative skills practiced during the teaching and learning process of practical mode in a non-face-to-face manner. The practical report rubric is an evaluation guideline to assess the practical reports prepared by students. Meanwhile, the formative test aims to test the students' mastery of Physics concepts applied during the practical of Simple Harmonic Motion.

2.1.1. Scientific Skills Checklist.

The Scientific Skills checklist was prepared by a group of researchers based on the Primary School Standard Curriculum (KSSR). The checklist is prepared using Google Form uploaded in Google Classroom a week before the scheduled class. Information related to Scientific Skills consisting of science process skills and manipulative skills was also uploaded as a guide to the participants while completing the scientific skills checklist before the PhET Simulation intervention was introduced to the participants. This Scientific Skills checklist was given to the participants before and after the PhET Simulation intervention was introduced.

2.1.2. Student Practical Report Rubric.

Rubric is an assessment tool to evaluate performance of the participants through practical reports provided. This rubric allows the researchers to assess the mastery of Scientific Skills by the participants. The rubric was prepared by the researchers based on the basic rubric guideline from the Matriculation Division of the Ministry of Education Malaysia. The rubric is then uploaded in the Google Classroom a week before the scheduled class, so that the participants are aware of the criteria being assessed in the practical report that will be prepared by the participants. The researchers reminded the students to read and understand the Practical Report Rubric, as well as to contact the supervising lecturer if there are any problems occurred related to the practical to be conducted in the following week.

2.1.3. Formative test

Formative testing is a strategy in the teaching and learning process, beneficial to help students and lecturers. In this action research, formative tests were given to the participants after their practical of using PhET simulations were implemented. The main purpose was to obtain an immediate feedback on the level of mastery of the participants. This formative test had 15 questions that tested the participants related to the objectives of the experiment, the theory of physics in the experiment, the apparatus used in the experiment, the experimental procedure, recording the experimental data and the experimental discussion.

2.2 Practical lesson

Researchers used PhET simulations during practical mode. There were no control groups in this action research. Table 1, shows the work steps or actions taken by the researcher during the course of this action study.

Table 1. Actions taken

No	Description / Activity	Date	Remarks
1	Description of action research		Using online face-to-face method through *Google Meet platform*
2	Scientific Skills Checklist		
3	PhET Simulation	17th January 2021	
4	Practical Report Rubric		Scientific Skills Checklist using *Google Form* and uploaded in the *Google Classroom.*
5	*Simple Harmonic Motion* Experiment	24th January 2021	The preferred usage of smartphone, tablet or laptop by the participants.
6	Preparation of Participants' practical reports.	24th January until 30th January 2021	
7	Scientific skills checklist		Scientific skills checklist using *Google Form* is uploaded through *Google Classroom..*
8	Formative test	31st January 2021	
9	Evaluation of practical reports	31st January until 14th February 2021	Supervised lecturer

3 Result and Discussion

3.1 Distribution of participants by gender

A total of 36 study participants were selected for this action research and were classified into male and female participants. The distribution of participants by gender is shown in Table 2. It was found that 8 (22.2%) participants were male and 28 (77.8%) were female. This means that female participants are more numerous than male participants. The difference between female and male participants was 20 people which is 55.6%.

Table 2. Distribution of participants by gender

Gender	Frequency	Percentage
Male	8	22.2
Female	28	77.8
Total	36	100.0

3.2 *Scientific Skills Checklist*

Descriptive analysis uses frequency and is expressed in percentages. Table 3 shows the frequency and percentage of Science Process Skills practiced by participants before using PhET simulation (Pre) and after using PhET simulation (Post). The Science Process Skills listed are based on the Primary School Standard Curriculum (KSSR) issued by the Ministry of Education Malaysia. The Science Process skills involved are observing, classifying, using numbers, predicting, communicating, interpreting data, controlling variable, and experimenting. Table 3 shows a significant improvement for each Science Process Skills practiced by the participants, for example for Science Process Skills controlling variables and experimenting showed a 100% increment, this shows that participants were able to practice these Science Process Skills by using PhET simulations even when they were not in a real laboratory.

Table 3. Checklist of participants' process science skills

Scientific Skills checklist	Pre		Post	
	Frequency	Percentage of frequency	Frequency	Percentage of frequency
Observation	18	50	35	97
Classifying	0	0	16	44
Using Numbers	36	100	36	100
Predicting	28	77	25	69
Communicating	21	58	25	69
Interpreting Data	26	72	36	100
Controlling variable	0	0	34	94
Experimenting	0	0	30	83

Table 4 shows the Manipulative Skills conducting experiments practiced by the participants before using the PhET simulation (Pre) and after the use of the PhET simulation (Post). Table 4 shows a clear improvement for sorting materials and apparatus of 94%, according to all the procedures specified to conduct 97% experiments and conducted the process of collecting data 97%. By using PhET simulation, participants must assemble their own virtual apparatus during the PhET simulation before they can obtain data.

Table 4. Manipulative skills checklist (doing experiment) for participants'

Participants' Manipulative skills Checklist while doing Experiment	Pre		Post	
	Frequency	Percentage of frequency	Frequency	Percentage of frequency
Arranging materials and apparatus	1	3	34	94
Follow all the procedures in conducting the experiment	9	25	35	97
Data collection process	18	50	35	97
Arrange and store back the apparatus	1	3	1	3

Table 5 shows the Manipulative Skills Checklist (Data Presentation) practiced by the participants before using the PhET simulation (Pre) and after the use of the PhET simulation (Post). Based on table 5, the percentage of study participants practicing Manipulative Skills (Data Presentation) did not show a clear difference, however, still showing improvement. This situation occurred because the data and graph assessment format were emphasized to the participants at the beginning of the 1st semester study session again.

Table 5. Manipulative skills checklist (data presentation) of participants'

Manipulative skills Checklist (Data assessment dan Graphs)	Pre		Post	
	Frequency	Percentage of frequency	Frequency	Percentage of frequency
Construct appropriate tables (columns, rows, headings, and units) to record data	31	86	36	100
Identify variables	32	89	36	100
Record data to the correct decimal place (consistent and constant)	31	86	36	100
Systematic, quality, and computational	31	86	36	100

Table 6 shows the Manipulative Skills Checklist (Data and Graph Assessment) practiced by the participants before using the PhET simulation (Pre) and after the use of the PhET simulation (Post). Based on table 6, the percentage of participants practicing Manipulative Skills of data assessment and graphs showed no difference. This situation occurred because the data and graph assessment format were emphasized to the participants at the beginning of the 1st semester study session again.

Table 6. Participants' manipulative checklist (data assessment and graph)

Manipulative Checklist (Data Assessment and Graph)	Pre		Post	
	Frequency	Percentage of frequency	Frequency	Percentage of frequency
Writing of title	36	100	36	100
Stating the axes and units	36	100	36	100
Using an even scale	36	100	36	100
Data plotting	36	100	36	100
Graph size 75% of graph paper	36	100	36	100
Draw the best lines	35	97	36	100

3.3 *Observation of Students' Practical Report.*

Meanwhile, Table 7 shows the Manipulative Skills Checklist (Practical Report Writing) practiced by the participants before using the PhET simulation (Pre) and after the use of the PhET simulation (Post). Based on table 7, the percentage of participants practicing Manipulative Skills of practical report writing did not show a clear difference, nevertheless, it still showed improvement. This situation occurred because the format of writing the Practical Report has been set in the Physics Laboratory Manual booklet for Semester I & II DP014 & DP024, 2018.

Table 7. Participants' Manipulative Checklist (Practical Report Writing)

Participants' Manipulative Checklist (Practical Report Writing)	Pre		Post	
	Frequency	Percentage of frequency	Frequency	Percentage of frequency
Objective	35	97	36	100
Theory	35	97	36	100
Apparatus Checklist	35	97	34	94
Procedure	36	100	36	100
Data scheduling	36	100	36	100
Data Analysis	35	97	36	100
Discussion	35	97	36	100
Conclusion	36	100	36	100

3.4 *Simple Harmonic Motion Formative Test.*

Table 8 shows the average Formative Test scores of participants for the Simple Harmonic Motion experiment. The total marks of the Formative Test are 15 marks. The average score of the study participants was 13.9 indicating a high level of mastery by the participants. This formative test has 15 questions which tested the participants related to the experimental objectives, the theory of physics in the experiment, the apparatus used in the experiment, the experimental procedure, recording the experimental data and the experimental discussion.

Table 8. Simple Harmonic Motion Formative Test

Participants	Average
36	13.9

3.5 Changes in Participants Behaviour.

The researcher found that all participants diligently prepared the Simple Harmonic Motion Practical Report. In addition, participants who were relatively passive in the classroom showed positive changes. They became more active in attempting to questions and answers session with the classmates based on the discussions of the study where participants used the WhatsApp group platform. Through the non-face-to-face teaching and learning mode, students were found to feel more comfortable learning and were not afraid to ask questions. They do not feel embarrassed when giving the wrong opinion. In addition, peers play an important role in mentoring other peers. The researcher through his observation in the WhatsApp group platform, found that the students had fun doing PhET simulation activities. They were able to process the experimental variables according to their understanding without fear. They were free to do experiment. However, in the end, the experimental findings do not deviate from the theory stated in the Physics Laboratory Manual Booklet for Semester I & II DP014 & DP024, 2018. The PhET simulations introduced by the researchers cannot only help students master Scientific Skills well but also have created a fun learning environment. Student-cantered strategies and materials combined in a virtual simulation is a cooperative learning method giving the role of facilitator to the researcher, facilitating classroom control, and subsequently improving the teaching and learning practices of the researcher in the subject of Physics.

3.6 Strengths and Weaknesses of PhET Simulation.

The strengths of PhET Simulations can provide many benefits as researchers doing demonstrations using real equipment in a Physics lab. Participants can do experiment, where actual equipment is not available or impractical to install. They can also use PhET Simulations to perform experiments that are not possible for example experiments involving electrons. Participants found it easy to change variables that were difficult or impossible to change with real tools. In addition, participants were able to run simulations on their own equipment such as their own laptops, smartphones, or tablets anywhere, either to repeat or extend the experiments to further strengthen their understanding. The success of the learning process using PhET Simulation is highly dependent on the self-efficiency of the participants, this is one of the weaknesses of PhET Simulation. There was no monitoring from stakeholders to monitor the teaching and learning process of the participants. The second disadvantage, to use PhET simulation is very dependent on the equipment that the participants have, or computer equipment provided by the college. A third drawback is participants who were not proficient or less proficient in the use of computers would feel very unfavourable to use PhET simulations.

3.7 Objectives Achievement and Effectiveness of Action Research.

The researcher found that all the objectives of the study studied have been proven and achieved by the marks obtained by students during the formative test, evaluation of practical reports and checklist of Systematic Skills of participants. Through this action research study all the objectives of the study have been answered in detail and successfully proven that the Teaching and Learning (T&L) by using PhET simulation has been able to strengthen the Scientific Skills of students.

4 Conclusion

The researcher will maintain the use of PhET simulation because the researchers think that based on the findings of the study, this PhET simulation is the easiest and most effective way in empowering the Scientific Skills of students. In fact, the exposure of various teaching and learning methods to students such as the use of effective methods is necessary so that they do not face problems when they are in unexpected situations, for example the situation of Covid-19 pandemic. The researchers found that this action research can be conducted over a longer period than now and can use the various simulations available in PhET to get an effective effect. Therefore, a longer period will make the study more rational, effective, and efficient in terms of obtaining efficacy and validity on the effectiveness of the use of these PhET simulations. The researchers hope that this method will be extended to other classes and to other subjects such as science, biology, chemistry, and others.

Acknowledgement
This project is funded under Fundamental Research Grant Scheme (FRGS) FRGS/1/2018/SSI09/UTM/02/5 by The Ministry of Education Malaysia with vote number R.J130000.7853.5f067.

References
[1] Kurikulum Standard Sekolah Rendah 2014 *No Title* (Putrajaya Selangor, Malaysia)
[2] Raj R G and Devi S N 2014 *Sch. Res. J. Interdiscip. Stud.* **2** 2435–2443
[3] Shahwaliah F 2014 *J. Penyelid. Kent* **13**
[4] *Physics Laboratory Manual Semester I & II DP014 & DP024 2018 Putrajaya* (Malaysia: Wilayah Persekutuan Putrajaya)
[5] Fadzil H M and Saat R M 2017 E *Eurasia J. Math. Sci. Technol. Educ.* **13** 4591–4607
[6] *(MQA) A K 2020 March 29 Panduan Pengendalian Program Pendidikan Tinggi Semasa Dan Pasca Perintah Kawalan Pergerakan Covid-19 Cyberjaya Selangor, Malaysia*
[7] Abdullah S B 2011 *J. Edupress* **1** 301–305
[8] Gamage K A, Silva E K D and Gunawardhana N 2020 *Educ. Sci.* **10** 301
[9] Fox M F, Werth A, Hoehn J R and Lewandowski H J 2020 Teaching labs during a pandemic: Lessons from Spring 2020 and an outlook for the future. arXiv preprint arXiv:2007.01271.
[10] Ray S and Srivastava S 2020 *J. proteins proteomics* **11** 77–80
[11] Astutik S and Prahani B K 2018 *Int. J. Instr.* **11** 410–424
[12] Wieman C E, Adams W K and Perkins K K 2008 *Science.* **322** 682–683
[13] Moore E B, Chamberlain J M, Parson R and Perkins K K 2014 *J. Chem. Educ.* **91** 1191–1197
[14] Harahap F, Nasution N E A and Manurung B 2019 *Int. J. Instr.* ***12*** 521-538
[15] Burnes B 2020 *J. Appl. Behav. Sci.* **56** 32–59
[16] Yaumi M 2016 *Action Research: Teori, model dan aplikasinya* (Jakarta: Prenada Media)

Students' understanding of electric fields in the context of multiple representation

N Rizkiyanti[1*] and J Mansyur[1]

[1]Sub-Department of Physics Education, Faculty of Teacher Training and Education, Universitas Tadulako, Jl. Soekarno Hatta KM. 9 Kampus Bumi Tadulako Tondo Palu-Central Sulawesi, Indonesia.

*Corresponding author: nikenriskianti@gmail.com

Abstract. This study aims to describe physics students' understanding of the concept of electric fields through a multi-representation review in a qualitative way. This research is a qualitative research using test and interview techniques to obtain the data.The research subjects consisted of 6 students from Physics Education department. The test used is an essay test which includes indicators of translation, interpretation and extrapolation on the concept of an electric field. The more in-depth interviews with respondents related to the concept of the electric field were carried out in this study. Based on data analysis, it can be concluded that students' conceptual understanding is in the moderate category. The highest scores obtained are in translation, interpretation, and extrapolation respectively, which show that the level of indicator complexity affects student achievement. This research implies that the poor conceptual understanding of students on extrapolation indicators caused by the lack of creativity in students' thinking in exploring the whole information presented on a particular issue. Thus, a learning method that priorities high-level critical thinking skills is extremely needed.

1. Introduction

Physics is closely related to natural phenomena in the universe, such as phenomena on electrically charged objects. In an electrically charged object, various important phenomena occur, such as the transfer of electric charge, electric force, or changes in electric potential energy. Every human being needs electricity to support daily activities. Without realizing it, the need for electrical energy will continue to increase along with the growth in population, technological developments, and education development. There is an awareness of the current mindset that the big challenges for society related to electronic devices and energy-efficient computing related to electricity require better-integrated technology [1]. Success in developing technology cannot be separated from assessment activities in understanding the concept of electricity. Understanding is the basis for students to build insight. Misunderstanding of physics concepts can create gaps between teaching and students' understanding and impact students' learning outcomes [2].

One of the most essential components in the material on electrical topic that needs to be investigated is the electric field. Many methods can be carried out to learn the concept of electric fields in physics. One of them is through multi-representation. Learning activities that involve multiple representations provide a rich context for students to comprehend a concept [3]. Students' success in solving physics problems is influenced by the ability to represent the algebraic-based physics material [4]. These representation formats are very essential in studying physics, including verbal representations, pictures, graphs and mathematics [5]. Graphical representation can help students to conceptualise abstract mathematical structures in vector calculus when understanding Maxwell's equations [6]. It is necessary to introduce learning models and teaching materials with multiple representations to students in order to gain a better understanding. The right representation choice can generate a simpler view of the dynamic mechanism by shifting some of the complexity into the representation itself [7]. Therefore, students' representational abilities make a significant contribution in understanding the concept of an electric field. However, there are also some difficulties encountered when using multiple representation formats in understanding physics concepts.

Students have difficulty representing electric field vectors using representational geometric features and solving electric field problems [8]. Students' understanding of the field concept in electromagnetism shows that most students do not have a coherent conception of electric field characteristics [9]. Many students failed to choose the appropriate coordinate system when writing mathematical expressions for vector fields to solve electric field problems [10]. Students still have difficulties in recognizing the representation of electric field lines compared to vector field plots, algebraic notation, and the density of field lines as the magnitude of the electric field [11].

Research conducted by Gire and Price on understanding the concept of electric field vectors only focuses on using mathematical representations [8]. Bollen et al. have conducted research on students' difficulties in symbolic representation of electric field materials [10]. In addition, Campos et al. have conducted an investigation on students' understanding of the concept of an electric field that focuses on mathematical representation [11].

The above studies only focused on the format of mathematical representations but have not conducted research using verbal representation, pictures and graphics format. Therefore, it is necessary to conduct a study on the understanding of electric field concepts in physics students through a multi-representation review using verbal representation, picture and graphics format.

2. Method

This research was descriptive qualitative. It describes physics students' understanding of the concept of electric fields through a multi-representation review in a qualitative way. The subjects of this study were 6 students of Physics Education department. The data were collected through an essay test given to the respondents. Then, the interview in which two respondents used as the representative of each category of high, moderate, low was carried out. The essay test instrument contains the concept o f an electric field which is presented in a multi-representation form. The essay test instrument used in this study consisted of 4 numbers, consisting of 3 aspects of representation: verbal, pictures and graphics. It also uses three indicators of concept understanding: translation, interpretation and extrapolation. After obtaining the data, qualitative data analysis was carried out using three stages: reducing data, presentation of data and drawing conclusions and verification. Descriptive analysis technique wass carried out by analysing the match between the answers and the reasons for the respondents' answers which can show the mastery of the concepts possessed by the respondents on the electric field topic.

3. Results and Discussion

3.1 Indicator of Translation

As an example, the answer to RT-01 is presented for the context of the translation indicator in describing the picture as follows:

> The electric field lines formed in the figure were attracted to each other because the electric charges are different from each other. It is also seen in the picture that the arrow lines on the positive charge point were oriented outward which means releasing electrons. Then, on the negative electric charge, the arrows on the lines were directed inward (towards the nucleus) which means it is receiving electrons released by a positive electric charge.

RT-01 explained that the electric field lines formed in the picture were attracted to each other because the electric charges on the two sides have different types. RT-01 can accurately describe the direction of the arrow as a sign of the direction of the electric field of the two charges, pointing from a positive charge to a negative charge. However, RT-01 did not describe the number of electric field lines formed between the two charges. This should be included in the answer presented by RT-01 because it is closely related to the electric field strength in terms of the value of the two electrical charges.

The limited explanation given by RT-01 in the interview is due to the respondent's lack of memory recall of the electric field concept and the answers given in the testing session. This phenomenon is in line with the findings of Sutriani and Mansyur that the poor memory of respondents at each stage of problem-solving provides a great opportunity for errors to occur [12].

3.2 Indicator of Interpretation

For example, the answer to RT-02 is presented for the context of the interpretation indicator in interpreting the graph as follows:

In the image, the graph is decreasing. Thus, noticeably that the electric field strength becomes weaker if the distance is increased.

RT-02 only gave the interpretation that the graph was decreasing, so that the electric field strength became weaker if the distance was increased. RT-02 only stated the field strength. He did not clearly state the particular electric field strength. RT-02 also did not interpret the dash line which should be able to provide important information related to the comparison between the electric field strength and the charge distance. When asked about the interpretation of the graph, RT-02 answered:

RT-02 : *So the graph is decreasing, as the graph becomes further,then, it will decrease. So the electric field strength becomes weaker if the distance is increased*

The researcher : *How can you tell that the electric field strength is inversely proportional to the distance, which part of the graph do you think contains that information?*

RT-02 : *That part of the curve.*

The researcher : *So, what do you think the dash line on the graph mean in this case?*

RT-02 *From the value of 0 to this R, the value of the field strength is zero*

Based on the interview result above, RT-02 explained that the further away the graph was, the lower it would be. So, the electric field strength would decrease as the distance was also increased. The answer to RT-02 was still the same as the answer when working on written essay questions. During the interview, RT-02 explained the meaning of the dash line on the graph that the value 0 to R had a field strength value of zero. Compared to the answers from by RT-02 when answering written questions, RT-02 gave more answers during interviews.

RR-01 was given two images. The figure (a) shows a positive electric charge, which transfers electric current to negative charge. Meanwhile, figure (b) has a negative electric charge, which receives current from a positive electric charge. RR-01 interpreted figure (a) as a charge that transferred current to a negative charge. It was perceived from the direction of the arrow in the Figure. In Figure (b), RR-01 interpreted that a negative electric charge received current from a positive charge. There is only one point charge in each image, but the RR-01 understands it as two interacting charges. Thus, this is the error made by the respondent.

Respondents' answers were further explored through the interview method. The transcription is described as follows:

The researcher : *What meaning do you catch based on the picture in the question?*

RT-01 : *So in picture A, it is a positive charge, while in picture B, it is a negative charge. The positive charge direction of the electric force goes outward because it gives a negative charge. Then, for the negative the direction of the arrow is inward because it receives charges or ions.*

Based on the interview above, it can be concluded that the interpretation of RS-01 is in line with the results of the answers when working on written essay questions. RR-01 explained that the positive charge seen from the direction of the arrow in the picture pointed to the outward, means that the direction of the electric force is out because it gives a negative charge in the form of charges or ions. While for a negative charge, it only accepted electrons, so that the arrow pointed to the center of the negative charge.

3.3 Extrapolation Indicator

As an example, RS-01's answer is presented for the context of extrapolating indicators in determining the type of charge and describing electric field lines as stated "The charge at point B is positive, the charge at point C is negative". RS-01 responded correctly that at B was a positive charge and at C was a negative charge. Unfortunately, the RS-01 could not properly describe the electric field lines formed on the charges at A and D. RS-01 illustrated the electric field lines at A will point to the charge at D and past the charges at B and C. The number of lines and the direction of the electric field lines depicted were also incorrect.

Respondents' answers were further explored through the interview method with the transcript described as follows:

The researcher	:	*What do you think should be in point B and C?*
RT-01	:	*The charge at point B is positive while the charge point C is negative*
The researcher	:	*Can you give an explanation related to the picture of the electric field lines formed at point A and D based on the charges at point B and C that you have described?*
RS-01	:	*I think the direction of the line from point A to D is mutually attractive because only one point is positive and the other is negative.*
The researcher	:	*So, do you mean that there is an interaction between charges at point A and D?*
RS-01	:	*Yes, they interact with each other because they are positive and negative.*

Based on the interview above, it can be concluded that RS-01 believed that there was an interaction between a positive charge at point A and a negative charge at point D. RS-01 ignored the charge in point B and C which should be considered in describing the electric field lines. The answer that RS-01 gave for the charge at point B was a positive charge and at point D was a negative charge. That was in line with the answers given when working on written questions.

3.4 Students' Understanding on Electric Field

There were several essential information as a finding in this study, including the respondent's understanding that if there are different charges, in this case, are positive and negative charges that interact with each other, the electric field lines formed will attract each other. The direction of the electric field lines formed starts from a positive charge towards a negative charge. Respondents understand that the further the distance between the charges, the weaker the electric field strength of the two charges. This was in line with the findings of Jelicic, Planinic, and Planinsic in their research that students consider the farther away from the center of charge, the effect of electromagnetic induction from the generated electric field will be weaker [13]. However, some findings contradict the electric field theory. RR-02 thought that the increasing distance from the charge center, the stronger the electric field would be. Reinforcing the findings of Mustakim et al, that respondents made conclusions by discussing conflicting data with the problems encountered [14]. In addition, the next finding in this study was that respondents understood the arrow image as electric field lines. In line with the findings of Bollen et al, in their research results, students were not aware of the difference between electric field lines and vectors. They did not fully understand the representation of electric fields [15].

The next interesting finding was in extrapolation indicators. Respondents only focused on the direction of the electric field lines directing from a positive charge to a negative charge. So, the electric field lines would pass through the two charges between the positive and negative charges. Confirming Campos' findings that students did not understand the density of electric field lines and converted it into the algebraic form [11], this study also found some respondents who did not pay attention to the density of electric field lines when asked to describe electric field lines. Based on the study results, it showed that the understanding of the concept of electric fields in the translational aspect was quite good. This was proved by the acquisition score for the translation aspect, obtaining the highest score compared to other aspects. Respondents can described the pictures in verbal form quite well. As research conducted by Cao and Brizuella obtained, students understood concepts more easily through representation [16].

In line with the research conducted by Eynde et al, the respondents in that research had a good conceptual understanding of the translational aspects of the concepts of algebra and calculus [17]. However, in that research, there were still some respondents who could not translate between graphical and symbolic representations of mathematical relationships. The findings of this study were different from the research findings obtained by Wu and Liu that students' conceptual understanding of the translation aspect was mostly poor. It was obviously explained through some of the respondents' answers in their research that many were still failed in representing tables and figures and tend to be wrong in representing tables in verbal form [18].

Based on the study results, it was found that students' understanding of the interpretation aspect as depicted from the test results and interview quotes showed that most of the respondents were quite good at interpreting the questions in the form of image and graphic representations into verbal form. Although some respondents did not interpret the answers completely, their responses regarding the relation between the electric field strength and the distance of its electric field were sometimes incomplete. There were even the answers that erroneously distinguished between electric field vectors and electric field lines for questions in the form of image representation.

In line with a study conducted by Cnevic et al, some students did not achieve maximum results in interpreting schematic representations [19]. Students tend not to understand the theory of light waves well, so they could not write complete answers. During the learning process, students tend to be interested in visual things only. Based on the study result, it was found that the students' understanding of the electric field concept in the extrapolation aspect, as seen from the test results and interview, was still poor. Most of the students were unable to estimate the type of charge in the picture. Therefore, the respondents could not conclude the direction of the electric field formed in the figure. It was seen that some students made a mistake in determining the type of charge. There was still a sprinkle of errors when describing electric field lines. In addition, many students ignore the number of electric field lines depicted. The inability of students to understand the problem is the main cause of student failure in solving a problem [20].

In line with the findings of Pladninic et al, most of the respondents in their research could not predict the symbols or slope patterns of the graphs presented in the questions, causing most of the respondents to give wrong answers [21]. Contrary to the research results obtained by Susac et al, it showed that most students possessed a good understanding of extrapolation aspects. This can be seen from the number of students who generally succeed in estimating the shape of the graph from other contexts in the form of mathematics on kinematics topic [22]

Overall, the results of this study indicated that if sorted from the lowest three aspects of understanding, the following conclusions are obtained respectively (1). Extrapolation, (2). Interpretation, (3). translation. It showed that there were different levels among each aspect of understanding in the electric field concept.

4. Conclusion

Based on the description above, conceptual understanding can be inferred as the ability to grasp information regarding the concept and other related information, then, can restate the concept into a more in-depth interpretation. Based on the results of the study, it was found that the conceptual understanding of the electric field in the extrapolation indicator was at the lowest level compared to the translation and interpretation indicators. The students' poor conceptual understanding on extrapolation indicators is caused by the lack of creativity in students' thinking in exploring the overall information presented on a particular problem. Thus, a learning method that prioritises high-level critical thinking skills is extremely needed.

It is expected that students should explore their translation, interpretation, and extrapolation skills in each practice session, especially, in representing the form of questions. Not only through representations in the form of images, but also in the form of graphics and verbal. Thus, it will increase students' understanding of the concept of electric fields.

Future researchers are expected to conduct further research on students' conceptual understanding of electric fields and how to overcome the problem. Since it is essential to master conceptual learning, thus it can impact students' ability to solve problems.

Acknowledgement

We highly acknowledge the support of the Program Kompetisi Kampus Merdeka (PK-KM) Grant of the Education, Culture, Research, and Technology of the Ministry of the Republic of Indonesia, with funding identity: 119/E1/KM.05.03/2021.

References

[1] Biswas B and Saha B 2019 Development of semiconducting ScN *Phys. Rev. Mater.* **3**
[2] Lee H and Schneider S E 2015 Using astronomical photographs to investigate misconceptions about galaxies and spectra: Question development for clicker use *Phys. Rev. Spec. Top. - Phys. Educ. Res.* **11**
[3] Izsák A and Sherin M G 2010 Exploring the Use of New Representations as a Resource for Teacher Learning *J. Sch. Sci. Math.* **103** 18–27
[4] Finkelstein N D, Adams W K, Keller C J, Kohl P B, Perkins K K, Podolefsky N S, Reid S and Lemaster R 2005 When learning about the real world is better done virtually: A study of substituting computer simulations for laboratory equipment *Phys. Rev. Spec. Top. - Phys. Educ. Res.* **1** 1–8
[5] Salkind G M 2007 *Mathematical Representations* (George Mason University)
[6] Bollen L, Van Kampen P and De Cock M 2015 Students' difficulties with vector calculus in electrodynamics *Phys. Rev. Spec. Top. - Phys. Educ. Res.* **11**
[7] Mitra A, Sola I R and Rabitz H 2008 Understanding the role of respresentation in controlled quantum-dynamical mechanism analysis *Phys. Rev. A - At. Mol. Opt. Phys.* **77**
[8] Gire E and Price E 2014 Arrows as anchors: An analysis of the material features of electric field vector arrows *Phys. Rev. Spec. Top. - Phys. Educ. Res.* **10**
[9] Zuza K, Van Kampen P, De Cock M, Kelly T and Guisasola J 2018 Introductory university physics students' understanding of some key characteristics of classical theory of the electromagnetic field *Phys. Rev. Phys. Educ. Res.* **14** 1–11
[10] Bollen L, Van Kampen P, Baily C, Kelly M and De Cock M 2017 Student difficulties regarding symbolic and graphical representations of vector fields *Phys. Rev. Phys. Educ. Res.* **13** 1–17
[11] Campos E, Zavala G, Zuza K and Guisasola J 2020 Students' understanding of the concept of the electric field through conversions of multiple representations *Phys. Rev. Phys. Educ. Res.* **16** 1–9
[12] Sutriani and Mansyur J 2021 The analysis of students' ability in solving physics problems using multiple representations *Journal of Physics: Conference Series* vol 1760
[13] Serway and Jewett 2010 *Fisika Untuk Sains Dan Teknik Buku 2 Edisi 6* (Jakarta: Salemba Teknika)
[14] Mustakim M, Mansyur J, Hatibe A, Rizal M and Kaharu S N 2020 Analysis of Students' Causal Reasoning in Physics Problem Solving *Journal of Physics: Conference Series* vol 1521
[15] Bollen L, Van Kampen P, Baily C and De Cock M 2016 Qualitative investigation into students' use of divergence and curl in electromagnetism *Phys. Rev. Phys. Educ. Res.* **12** 1–14
[16] Cao Y and Brizuela B M 2016 High school students' representations and understandings of electric fields *Phys. Rev. Phys. Educ. Res.* **12** 1–19
[17] Van Den Eynde S, Van Kampen P, Van Dooren W and De Cock M 2019 Translating between graphs and equations: The influence of context, direction of translation, and function type *Phys. Rev. Phys. Educ. Res.* **15** 1–13
[18] Wu C J and Liu C Y 2021 Eye-movement study of high- And low-prior-knowledge students' scientific argumentations with multiple representations *Phys. Rev. Phys. Educ. Res.* **17** 1–16
[19] Matejak Cvenic K, Ivanjek L, Planinic M, Jelicic K, Susac A and Hopf M 2021 Analyzing high school students' reasoning about polarization of light *Phys. Rev. Phys. Educ. Res.* **17** 1–16
[20] Sardi S, Rizal M and Mansyur J 2018 Behaviour of mathematics and physics students in solving problem of Vector-Physics context *Journal of Physics: Conference Series* vol 1006
[21] Planinic M, Ivanjek L, Susac A and Milin-Sipus Z 2013 Comparison of university students' understanding of graphs in different contexts *Phys. Rev. Spec. Top. - Phys. Educ. Res.* **9** 1–9
[22] Susac A, Bubic A, Martinjak P, Planinic M and Palmovic M 2017 Graphical representations of

data improve student understanding of measurement and uncertainty: An eye-tracking study *Phys. Rev. Phys. Educ. Res.* **13** 1–20

Analysis of physics learning media needs based on mobile augmented reality (AR) on global warming for high school students

M. Nor[1,2*], and Lilia Halim[2]

[1]Department of Physics Education, Faculty of Teacher Training and Education, Universitas Riau, Riau, Indonesia
[2]Faculty of Education, Universiti Kebangsaan Malaysia, Selangor, Malaysia

*Corresponding Author: m.nor@lecturer.unri.ac.id

Abstract. The rapid progress of the times requires an educational approach to be innovative and adaptive to changes in technology and information. Mobile Augmented Reality (MAR) is a system that incorporates technology in education. This method facilitates learning to be interesting, effective, and easily accessible to everyone. This survey research was an initial step in developing Augmented Reality based Modules for high school students on global warming. The importance of this research is that educational media using Augmented Reality is relatively new in the field of educational media. Analysis of learning problems and analysis of students are the two indicators used in the analysis of this need. Based on this research survey, it was found that augmented reality based educational media is viewed as important and necessary as learning materials in schools especially for students of high ability. The next phase is the development of MAR-based learning media that meets these requirements.

1. Introduction

The development is so rapid in this world, so fast that Indonesia has achieved the Fourth Industrial Revolution. The advancement and ease of technology today have a significant impact on the lives of every Indonesian citizen including the education sector, are affected by [1]. In the world of education, there are important things that affect the learning process, which lies in the learning activities carried out by educators. Teachers should actively involve students in every learning activity during the learning process. In line with the current educational paradigm, namely the student centered method, which expects students to be more active participants in learning activities, this is appropriate [2].

Students have difficulty understanding physics learning, according to research conducted by [3], because the information is provided only through textbooks and only applies to problems involving formulas. Lubrecht [4], a teaching and learning process requires communication between two parties to create a fun and challenging environment where students can express themselves creatively, allowing their talents and interests to develop and maturity and independence to develop. Born to students. As a result, physics learning must be accompanied by appropriate learning tools to be effective [5]. In today's world, one of the most popular sources of learning is the utilization of learning media. Conversely, the current learning media is less able to visualize abstract processes, and has inadequate or excessive dimensions, resulting in difficulties in practice and direct observation, especially in terms of physical matter [6].

Print media, both textbooks, and modules are one of the learning media available. Arsyad [7] Teaching book media has the advantage that students can repeat the material without quickly tiring the eyes, follow the order of the mind rationally, and progress according to the speed of each student. While print media has the disadvantage that is difficult to convey movement on the page, it also has shortcomings that if not designed properly can cause students to become bored. In addition, the use of textbooks severely limits the learning opportunities available to students.

Supatro et al [8] learning media is always evolving along with technical advances, including print technology, audiovisual technology, computer technology, and a combination of print and computer technology. By technological advances, a book media format can be created that incorporates augmented reality (AR). This has the potential to reduce the shortcomings of the teaching book medium or module.

Based on the background information above, researchers are interested in analyzing the needs of mobile augmented reality (MAR) based physics learning media for high school students, with the title "analysis of the needs of mobile augmented reality (MAR) a global warming material based physics learning medium for high school students". Students need for offline and online teaching and learning processes is considered to be identified and mapped through the use of this research based analysis.

2. Method

Researchers surveyed at Senior High School SMAN 6 Pekanbaru. And SMAN 2 Tambang, Riau with 60 students survey respondents. The research was conducted in April 2021. The type of research conducted is survey research using online questionnaires (Google form). The flow of the research process is depicted in the diagram in Figure 1.

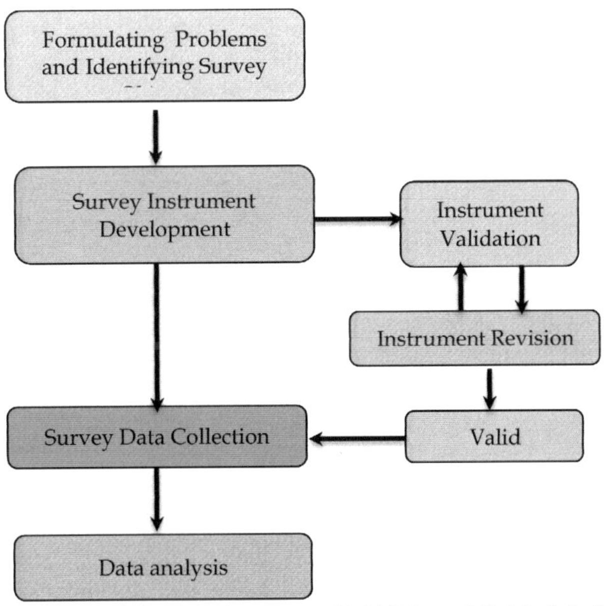

Figure 1. Flowchart of the development of MAR based E-Module learning tools

The questionnaire used in this study was prepared in response to the needs analysis indicators used in this study. The validity and reliability of this research tool have been determined. The questionnaire consists of question items derived from two indicators, as shown in Table 1.

Table 1. Research Instrument Statement Indicator

Analysis Indicators	Number of Question Items
Student Analysis	6
Learning Problem Analysis	4

The results of this study are used to assess the level of need, as shown in the following Table 2.

Table 2. Purposes Analysis Category

Category	Rank	Percentage	Decision
Strongly Disagree	1− ≤ 1,75	<16% - 37%	Very No Need
Disagree	>1,75 − ≤ 2,5	<37% - 58%	No Need
Agree	>2,5 − ≤ 3,25	<58% - 79%	Need
Strongly Agree	>3,25 - 4	<79% - 100%	Very Need

The answer choices for the question items are as follows: strongly agree with a value of 4, agree with a value of 3, disagree with a value of 2, and strongly disagree with a value of 1. To determine whether the questionnaire is appropriate or not. feasible, it will be validated by a professional validator before it is available to participants.

3. Results and Discussion

Dudley-Evans et al [9] in the analysis of needs that are the subject of research, there are three concepts of needs analysis, namely to know the condition of students, to know how learning can be maximized in a particular goal. conditions and groups, as well as to know the target situation and environment so that data can be interpreted appropriately. It is also supported by [10], who argue that this analytical model should consist of three subcategories of target needs analysis, among other needs, deficits, and wants. This is because, according to [11], learning needs are defined as tasks that students must complete before they are deemed to have acquired the necessary information and abilities. To that end, when defining or analyzing those needs, everything related to students should be considered, including their interests, talents, and background.

Kuo [12] needs analysis is a strategy to collect data and evaluate important information connected with research design to map the problem to be studied. learning e-module for high school students, this project conducts a needs analysis. In addition, this research is expected to be a starting point for the development of learning innovations that incorporate technology into education. As a result, using MAR based media, e-Module learning, which was previously unattractive, becomes much more successful. Teachers will find it easier to teach with this MAR based e-module learning, which will be assisted by the role of technology in the process. Similarly, students will be more interested in videos using 3D scanning modules and mobile phones because they will see virtual media as real. Learning takes place in an environment that relies heavily on technology integration. MAR agrees with this media learning. The results of this study will develop MAR based e-module learning for high school students, and two indicators will be measured and studied as a reference to determine the requirements for building MAR based e-module learning. There are four question items in the problem analysis indicators, including what learning methods have students experienced so far? what learning methods do students want? and what learning methods have students experienced so far? what learning approach suits the needs of students, and whether or not the method is effective in meeting those needs. For question item 1, which asks about what learning strategies have been used in e-module learning, Respondents provide answers as depicted in Figure 2. The following questions are intended to determine the effectiveness of the most frequently used approaches and the preferred learning methods of students in the learning environment. virtual. numbers 2 and 3 whose results are depicted in Figure 2.

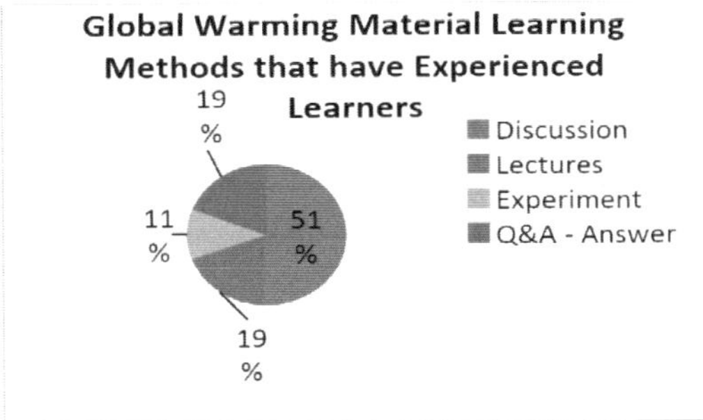

Figure 2. Learning methods that have been used so far

However, there are still some drawbacks and drawbacks to the discussion approach, which is the most widely used of the available methods. According to the statement, three learning strategies based on the discussion method have been identified as very helpful for learning physics. However, because learning can take place offline or online, the conversation must be packaged as a virtual or virtual world dialogue through the use of technology. To achieve this, learning must be accompanied by interesting and appropriate media resources. However, much has happened so far, the media used are less attractive and less effective. Based on the distribution of the survey in SMA 6 Pekanbaru in Riau Province, the results show that high school students need technology to understand physics courses on global warming which are included in the "agree" group. This can be seen in the following statement 4 in Figure 3:

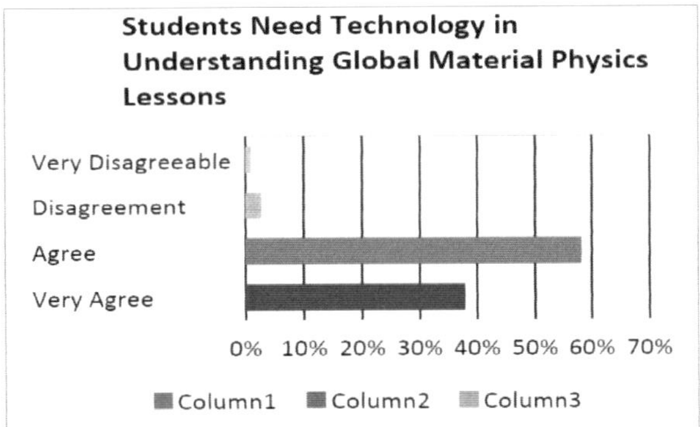

Figure 3. Questionnaire results about the problem less interesting e-module learning media and the use of technology in understanding physics learning

After analyzing the learning problems of the four questions posed in the questionnaire, it was determined that the most significant problem for students in e-module learning was the ineffectiveness and attractiveness of the media that had been used so far. For example, the media used is not interactive, does not have many characteristics of learning content, and does not allow students to complete the material in learning easily and effectively. Without improvised media, this makes the learning methods used ineffective, especially in terms of letting students learn through modules without the media. This also causes students to become less interested in actively participating in the learning that takes place and is taught by the teacher as a result of this. The right answer must be found for this problem; This solution is communication through media that is effective and interesting for students. In this example case, MAR. A research project directed by [13] investigated the effect of augmented reality-based experiments in increasing students' knowledge and cognitive load when compared to 3D and traditional trials.

There are a total of six question items in the student analysis. To understand the physics lesson on global warming information in question number 1, students must have access to technology. Question 2 resulted in the results that all students (60 percent) did not know Augmented Reality technology, and (40 percent) who filled out active questionnaires and often used smartphones in learning but did not use modules with MAR technology themselves; The survey results for the percentage can be seen in Figure 4.

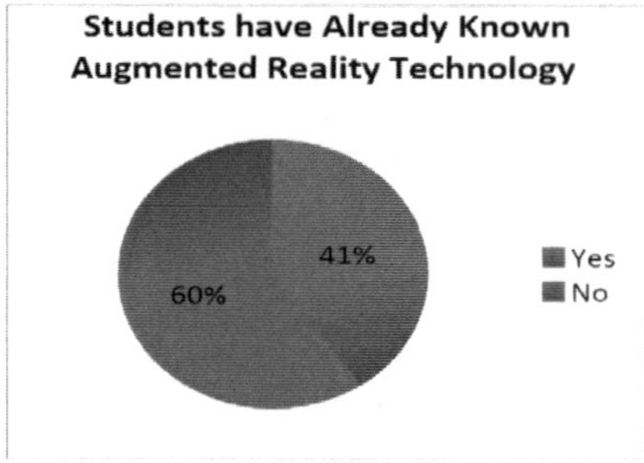

Figure 4. Students are already familiar with Augmented Reality technology

Based on the picture above, there are still most people who are new to Augmented Reality technology, but there are also those who are not familiar with MAR technology in understanding physical learning to improve the quality of learning and educational procedures.

This student analysis also examines students' requirements in offline and online learning environments, to ensure that teaching and learning activities remain communicative, successful, and enjoyable even if done digitally. Some questions were asked regarding the needs of students, including the use of smartphones with module connections to make learning more exciting and fun, as well as about students' requirements for effective and efficient media and learning materials in physical learning. In the following steps, the findings of the student need analysis survey are obtained, as illustrated in Figure 5:

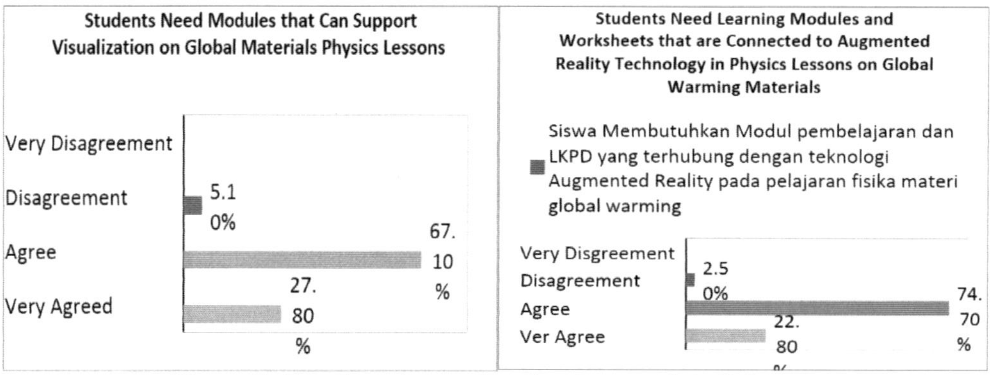

Figure 5. Questionnaire results of student needs in E-learning

Students agree with the choice stating that physics learning will be more exciting and enjoyable if transferred and integrated with a mobile-based physics learning system of global warming materials designed to help students understand abstract learning concepts, and developed modules can be used to help students understand abstract learning concepts in the survey results shown in the image above. In addition, it can also be an alternative media that teachers can use to teach more interestingly and interestingly. Students motivation to learn about global warming can be enhanced by the use of integrated, communicative, effective, engaging, and simple media to apply, as this medium shows. As a result, it generates greater interest and participation from students in learning and teaching activities while still ensuring that learning takes place successfully, both offline and online. It is intended to be used in this example medium which is based on the MAR technology system. Hendriyani [14] augmented reality technology can create any 3D model that may be difficult to visualize in the classroom, on a computer, or in a student's brain. Students gain a deeper understanding of the subject matter learned when abstract content is included in a visual model.

The needs analysis carried out in this study can be used as a reference and standard for current educational needs and needs. Students have a strong interest in the needs of MAR based media and learning tools as they are communicative, integrative, and effective media with a wide range of interesting features and materials. This is seen from the description of the results of the analysis of survey research data above. This advantage has the potential to be a significant source of capital to maintain the quality of education and increase the attractiveness of learning. In general, this MAR system is a very unique software that contributes to the learning environment by providing various advantages, among others [15]. Given the survey findings of each question item about problem analysis and student needs analysis described above, it can be determined that high school students, particularly those studying physics, have an urgent need for MAR based e-modules. Study in those subjects. For the survey results to answer the demands identified during the study, the survey must be completed.

4. Conclusion

The basic Mobile Augmented Reality (MAR) e-module needs analysis for high school students consists of two indicators, namely problem analysis, and student needs. With the findings of the study, there should be an e-module in MAR learning for high school students with a very agreeable and much-needed category level. As a result, the next stage is designing MAR media and the development of learning devices that fit current needs.

References

[1] Estheriani N G N and Muhid A 2020 *Insight J. Ilm. Psikol.* **22** 118–129
[2] Alfika A 2018 *Prosiding Seminar Nasional Fisika dan Pendidikan Fisika Universitas Ahmad Dahlan Yogyakarta. ISSN : 2477-1511*
[3] Hafi N N and Supardiyono S 2018 *Inov. Pendidik. Fis.* **7** 306–310
[4] Lubrecht A 2012 Augmented reality for education "Digital Union" The Ohio State University. (available at: http://digitalunion.osu.edu/2012/04/24/augmented-reality-for-education/)
[5] Chaeranti S N, Bakri F and Permana A H 2018 *Prosiding Seminar Nasional Fisika*, 7,118–128
[6] Listiaji P, H. M, Sugiyanto and H S 2019 *WaPFi (Wahana Pendidik. Fis.* **4** 216–223
[7] Arsyad A 2011 *Media pembelajaran* (Jakarta: PT. Raja Grafindo Persada)
[8] Saputro R E and Saputra D I S 2015 *J. Buana Inform.* **6** 153–162
[9] Dudley-Evans T and St John M J 2015 *J. Educ. Res.* **03** 24–31
[10] Hutchinson T and Waters A 1987 *English for specific purposes* (Cambridge university press)
[11] Alqunayeer H S and Zamir S 1987 *English for specific purpose* (New York: Cambridge University)
[12] Kuo S M 2016 *J. Appl. Linguist. Lang. Res.* **3** 36–53
[13] Liu Q, Yu S, Chen W, Wang Q and Xu S 2021 *J. Comput. Assist. Learn.* **37** 645–656
[14] Hendriyani Y, Effendi H, Novaliendry D and Effendi H 2019 A *J. Teknol. Inf. dan Pendidik.* **12** 62–67
[15] James G and Mazoué J G 2012 *J. Comput. High. Educ.* **24** 74–95

Identifying students' difficulty in the basic of thermodynamics

M C Sutarja[1*] and A Y R Wulandari[1]

[1]Department of Science Education, Faculty of Education, University of Trunojoyo Madura, Madura, Indonesia
*Corresponding author: maria.sutarja@trunojoyo.ac.id

Abstract. Conceptual understanding is one of the main topics in physics educational research. It is important as the basic of other ability in education, such as argumentation, making scientific explanation and problem solving. This survey study aims to identify students' difficulty to understand the basic of thermodynamics. There are 128 first year undergraduate students as the participants of the study. The data collection method is test. Result of the study shows that isobaric process of ideal gas and mechanical equilibrium state concepts are the most difficult concepts. Some difficulties are found in understanding the basic of thermodynamics: (1) because of the presence of higher cognitive load while solving conceptual problems, (2) when the question demands other ability, especially mathematical ability, to solve the conceptual problem, (3) because of students' disability to integrate the knowledge. This study could be used to develop learning instruction or media in basic physics or introduction of thermodynamics course.

1. Introduction

Students' conceptual understanding is one of the important topics in physics educational research. It is one of the main goals in physics learning. Understanding the concepts is important as the basic of other ability in learning, such as argumentation [1], making scientific explanation [2], and problem solving [3]. Describing students' understanding in physics is important to identify their difficulty then design learning to overcome it.

Some standardized tests have been developed to identify students' conceptual understanding and reveal students' difficulty, such as TCS [4], STPFaSL [5], TCE [6], and HTCE [7]. Many of them are designed in daily live context. Many studies used those instruments to describe students' conceptual understanding.

Thermodynamics is one of physics branch deals with heat, work, temperature, and energy. Students' understanding in this physics branch has been identified. Some of them focus on heat and temperature concepts and some other are on thermodynamics law [8][9][10][11]. Some of them reveal students' difficulty and some others reveal students' disability to think as the way of scientists think about the concepts (students were confused between their naïve understanding and scientific concept). Most students were in low level of conceptual understanding in thermodynamics, although they have been studied thermodynamics since their primary school [12]. They were low in some concepts such as relation between energy and enthalpy reactions, heat and temperature [9][11], first law of thermodynamics [13], and mechanical equilibrium principle [8]. The study about students' conceptual understanding continues, students' difficulty in understanding the concepts have been identified, and many interventions have been implemented to overcome many obstacles in education [3][14][15]. However, the data show that study to identify students' difficulty is still needed, because students' understanding differs based on their background and experience [16], especially during COVID-19 pandemic.

During COVID-19 pandemic, students face different learning atmosphere experience. This study held to identify students' difficulty in understanding the basic of thermodynamics concepts, especially during COVID-19 pandemic. The result of this study could be the base to design learning effectively or develop media to learn about the basic of thermodynamics.

2. Method

This was a survey study. There were 128 undergraduate science education students of first year as the participants of the study. The students have not studied the basic of thermodynamics in basic physics or thermodynamics course. Students have to answer questions in the survey study at the beginning of the course, before learning about the basic of thermodynamics. As undergraduate students, they have learned the basic of thermodynamics in high school.

Due to COVID-19 Pandemic, this study held online. The questions were presented in *google form* platform. The answer acceptance was limited based on the set time. There were 12 questions in bahasa (Indonesian language) as a test instrument. The instrument consisted of 3 topics is presented in Table 1. Eleven questions were taken from TCS [4], while 1 question was arranged to compare with one of the topics in the study. The instrument was statistically valid and reliable, with the number of reliabilities 0.69, analyzed using KR 20. The results of the test are analyzed descriptively.

Table 1. Categories of Questions

Topic of Questions	Sub-topic of Questions	Question Number in TCS	Question Number in study instrument
Heat and temperature	Thermal equilibrium	1 and 7	Q_1, Q_4, Q_5
	Heat transfer	5 and 6	Q_2, Q_3
Isobaric process of ideal gas	Isobaric process of ideal gas	9, 12, and 13	Q_6, Q_7, Q_8
Adiabatic	Adiabatic	14, 15, 16, and 17	Q_9, Q_{10}, Q_{11}, Q_{12}

3. Results and Discussion

Students' correct answer percentage are presented in Figure 1. Figure 1 shows that the most difficult topic is isobaric process of ideal gas, among topics asked in survey study. Only 29.69% of students answered the question in isobaric process of ideal gas correctly. Students' answer distributions are presented in Table 2, 3, and 4.

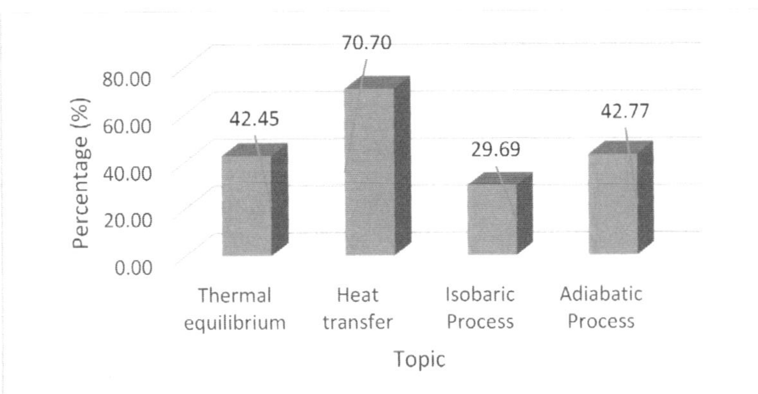

Figure 1. Percentage of students' correct answer in every topics

Thermal equilibrium topic was asked in the first, fourth, and fifth questions. Most students answer correctly in the fourth question, but less in the first and fifth questions. In the first question, the students were asked about the range of final temperature; in the fourth question, the students were asked to compare the heat transfer of the different volume objects; and in the fifth question, the students were asked about the exact number of final temperatures in thermal equilibrium state.

In the first and fifth questions, students needed mathematical ability to answer those two questions. Students answer distribution is presented in Table 2. The number of students' correct answer in both questions are similar, 46 and 48, although the first question asked only about the range of final temperature, in contrast to the first question that asked the students to state the exact number of final temperatures. Due to the low number of students' correct answer, it indicates that students were getting trouble in the questions asking about the final temperature in thermal equilibrium state, needed mathematical ability to solve it.

Students were asked about the same topic (rate of heat transfer between objects with different material) in the second and third questions, but the number of students answered correctly is quite different. 113 students answered correctly in the second question, but only 68 students in the third question, as presented in Table 2. The second and third questions are presented in Figure 2.

Jan announces that she does not like sitting on the metal chairs in the room because "when touching it, they are colder than the plastic ones." Which statement do you strongly agree with?

A) Jim agrees and says, "The metal chairs feel colder because metal is naturally colder than plastic."

B) Kip says, "The metal chairs are not colder because they are at the same temperature."

C) Lou says, "The metal chairs are not colder, the metal ones just feel colder because they are heavier."

D) Mai says, "The metal chairs are colder because metal absorbs the heat from body faster."

Kim picks up two rulers, a metal one and a wooden one. He announces that the metal one feels colder than the wooden one. What is your preferred explanation for this situation to Kim?

A) Metal conducts heat faster than wood.

B) Wood is naturally a warmer substance than metal.

C) Metals are better heat radiators than wood.

D) Cold flows more readily from a metal.

Figure 2. The second and third questions [4]

Table 2. Students' answer distribution of Thermal equilibrium and Heat transfer topics

Choice	Q_1	Q_2	Q_3	Q_4	Q_5
A	25	11	68	19	48
B	20	1	7	69	23
C	46	3	47	40	32
D	33	113	6	-	25
E	4	-	-	-	-

The correct answer

The fourth and fifth questions asked about heat transfer of objects with different materials. In the fourth question, students were asked to compare the rate of heat transfer between plastics and metal, while in the fifth question, between wood and metal. There were 113 students answer correctly in the fourth question, but only 68 students answer the fifth question correctly, as presented in Table 2. In the fifth question, most students choose C option as the answer. They chose the answer, failed in identification of heat transfer in metal. Choosing that option indicates that students don't have a good knowledge about heat transfer process. The fifth question is more complex than the fourth question, because the students need to have more and better knowledge to answer it.

The sixth, seventh, and eighth questions asked about isobaric process of ideal gas, as shown in Figure 3. Most students answer correctly in the sixth but less in the eighth question, as presented in Table 3. The sixth question asked about the change of ideal gas temperature in a syringe covered by frictionless movable piston with mass M. The syringe is moved from the vessel with cold water to the vessel with hot water.

Table 3. Students' answer distribution of isobaric process of ideal gas

Choice	Q_6	Q_7	Q_8
A	68	49	39
B	43	51	71
C	17	28	18

The correct answer

The seventh and eighth questions show three identical cylinders contain ideal gas. All of them are covered with frictionless movable piston with mas M. Cylinder A and B are in thermal equilibrium with 20^{0}C of room temperature, while cylinder C is kept to have 80^{0}C of temperature. All pistons are in mechanical equilibrium with the environment. The seventh question asked about the pressure of cylinder A compare to cylinder B, while the eighth question comparing B and C. The sixth, seventh, and eighth question are presented in Figure 3.

Please use the following information to answer question 6
A syringe that contains an ideal gas and has a frictionless piston of mass M is moved from a beaker of cold water to a beaker of hot water. Answer the following questions and consider that the syringe reaches thermal equilibrium with hot water.

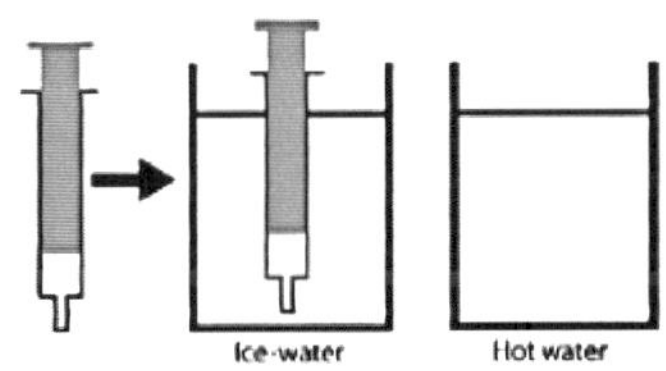

6 How does the gas temperature change?
A) Increase B) Decrease C) No change

Please use the following information to answer questions 7-8
Three identical cylinders are filled with unknown quantities of ideal gases. The cylinders are closed with identical frictionless pistons of mass M. Cylinder A and B are in thermal equilibrium with the room at 20°C, and cylinder C is kept at a temperature of 80°C. The piston of each cylinder is in mechanical equilibrium with the environment.

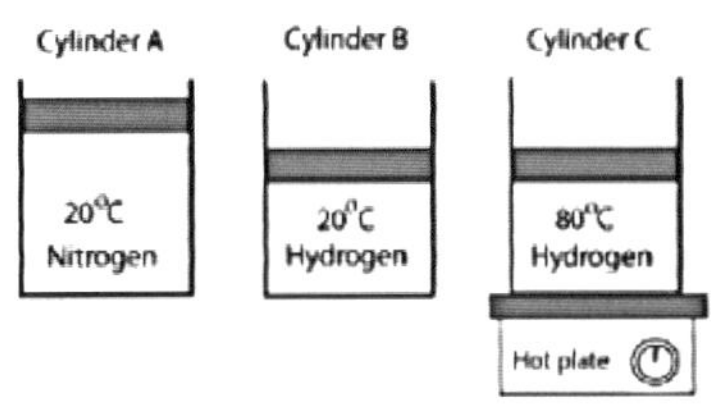

7 . How does the pressure of nitrogen gas in cylinder A compare with the pressure of hydrogen gas in cylinder B?
A) Greater B) Less than C) Same

8 How does the pressure of hydrogen gas in cylinder B compare with the pressure of hydrogen gas in cylinder C?
A) Greater B) Less than C) Same

Figure 3. The sixth, seventh, and eigth questions [4]

In both questions (the seventh and eighth questions) only a little number of students answer the questions correctly. The students couldn't think that the frictionless identical piston covering the cylinder can move freely due to the volume expansion of gas, to keep the gas in cylinder to have the same pressure (isobaric)

The case mentioned indicates that most students also fail to recall the concept of mechanical equilibrium state to answer those two questions. The pressure of ideal gas in cylinder is controlled by the pressure of environment outside the cylinder [17]. Instead of using this concept, students preferred to recall the concept 'higher temperature increase the pressure of ideal gas in cylinder' (based on 71 students choose 'smaller' when they were asked the gas pressure in cylinder B compare to cylinder C). This finding is also found in the previous study, that most students were getting trouble answering the question that involve mechanical equilibrium state and isobaric concept [8].

Mechanical equilibrium is a concept learned in the previous course. Students have studied this concept in basic physics course I. Students' disability to recall this concept while answering conceptual question about isobaric process of ideal gas indicates their disability to integrate one concept to another concept in physics, such as finding in previous study of momentum concept [18]. Their conceptual understanding was fragmented. When they learned about isobaric process of ideal gas, they left their understanding in a previous course. It happened because they think physics as fragmented, not as one body knowledge.

Due to the characteristics of those three questions, the seventh and eighth questions are more complex than the sixth question. These two questions also become the questions with smallest number of correct answers in this study. It shows that isobaric topic is the most difficult topics among thermal equilibrium, heat and transfer, isobaric, and adiabatic topics. Students' ability to answer the question degrade while solving the more complex question. It happened due to the presence of higher cognitive load [19]. This is the same as the finding in the second and third questions, when students fail to answer the question correctly due to the complexity of the question (more knowledge was needed to find the right answer).

The nineth, tenth and eleventh questions asked about the temperature (Q_9), pressure (Q_{10}), and volume (Q_{11}) change of ideal gas in isolation state when some objects are placed on the piston covering the syringe of this ideal gas, in Figure 4. The twelveth question asked about the change of 1 mole ideal gas temperature when the gas pushed with frictionless piston. Most students answer correctly in those three questions (the nineth, tenth and eleventh questions), as presented in Table 4.

The largest number of students answer correctly in the tenth question. The students could predict that the temperature and pressure are going to be higher when some objects are placed on the frictionless piston. Some of them, as the second largest number of students answer correctly in this topic, predict correctly that the volume is going to decrease when some objects are placed on the frictionless piston.

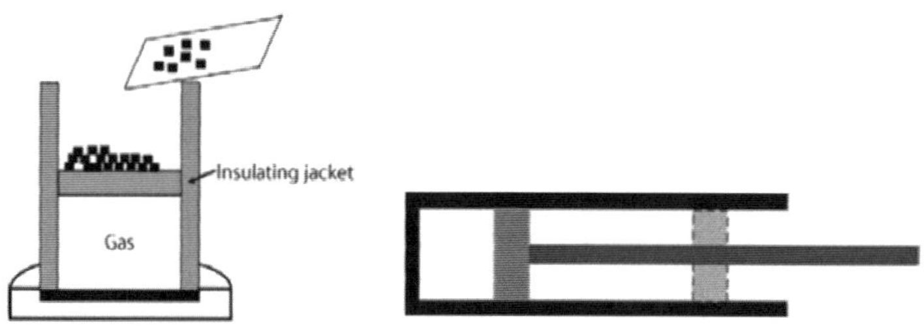

Figure 4. The Figure in the eighth-eleventh (left) and twelfth (right) questions [4]

Table 4. Students' answer distribution of adiabatic topic

Choice	Q_9	Q_{10}	Q_{11}	Q_{12}
A	46	66	29	61
B	37	30	46	38
C	45	32	53	29

The correct answer

4. Conclusions

Based on the result of this study, Isobaric of ideal gas is the most difficult topics among heat and temperature, isobaric process of ideal gas, and adiabatic process of ideal gas topics. The students' fail in recalling frictionless movable piston concept while answering the questions. This fail refers to the presence of higher cognitive load, in more complex question, such as more variables involve in a question that need to be solved. The students also fail in answering question correctly while it need mathematical ability to solve it. The result of this study also reveals students' disability to integrate one concept to another concepts. It indicates that students see physics as fragmented knowledge, not as one body knowledge.

The study was limited to identify students' difficulty in basic of thermodynamics. It didn't explain why students experienced this difficulty, especially during this COVID-19 pandemic. This result of study should be used to develop learning instruction or media in thermodynamics, to overcome students' difficulties in thermodynamics.

References

[1] Ogan-Bekiroglu F and Eskin H 2012 Examination of The Relationship Between Engagement in Scientific Argumentation and Conceptual Knowledge *Int. J. Sci. Math. Educ.* **10** 1415–43

[2] Maria Chandra Sutarja, Sutopo and Eny Latifah 2017 Hubungan antara Penguasaan Konsep dengan Kemampuan Membuat Penjelasan Ilmiah Siswa pada Topik Fluida Statis *Prosiding Seminar Nasional Fisika dan Pembelajarannya 2017* ed U N M Jurusan Fisika, Fakultas MIPA (Malang) pp 154–62

[3] Docktor J L, Strand N E, Mestre J P and Ross B H 2015 Conceptual Problem Solving in High School Physics *Phys. Rev. Spec. Top. - Phys. Educ. Res.* **11** 1–13

[4] Wattanakasiwich P, Taleab P, Sharma M D and Johnston I D 2013 Development and implementation of a conceptual survey in thermodynamics *Int. J. Innov. Sci. Math. Educ.* **21** 29–53

[5] Brown B R 2015 Developing and Assessing Research-Based Tools for Teaching Quantum Mechanics and Thermodynamics

[6] Chu H-E, Treagust D F, Yeo S and Zadnik M 2012 Evaluation of Students' Understanding of Thermal Concepts in Everyday Contexts *Int. J. Sci. Educ.* **34** 1509–34

[7] Tanahoung C, Chitaree R, Soankwan C, Sharma M and Johnston I 2006 Surveying Thai and Sydney Introductory Physics Students ' Understandings of Heat and Temperature *Proc. Assess. Sci. Teach. Learn. Symp.*

[8] Azizah D N, Sutopo and Zulaikah S 2017 Identifikasi Pemahaman Konsep Siswa SMA pada Materi Termodinamika *Pros. Semin. Pendidik. IPA Pascasarj. UM* **2** 134–42

[9] Budiarti I, Suparmi S, Sarwanto S and Harjana S 2016 Students'Conceptual Understanding Consistency of Heat and Temperature *J. Phys. Conf. Ser.* **755** 0–5

[10] Linuwih A M S, Fisika J, Matematika F, Alam P and Negeri U 2013 Konsepsi Alternatif Mahasiswa Fisika Pada Materi Termodinamika *UPEJ (Unnes Phys. Educ. Journal)* **2**

[11] Nugraha D A, Suparmi A, Winarni R and Suciati 2017 Students' Profile of Heat and Temperature Using HTCE in Undergraduate Physics *J. Phys. Conf. Ser.* **909** 0–6

[12] Saricayir H, Ay S, Comek A, Cansiz G and Uce M 2016 Determining Students' Conceptual Understanding Level of Thermodynamics *J. Educ. Train. Stud.* **4** 69–79

[13] Kulkarni V D and Tambade P S 2013 Assessing the Conceptual Understanding about Heat and Thermodynamics at Undergraduate Level *Eur. J. Phys. Educ.* **4** 9–16

[14] Nie Y, Xiao Y, Fritchman J C, Liu Q, Han J, Xiong J and Bao L 2019 Teaching towards knowledge integration in learning force and motion *Int. J. Sci. Educ.* **41** 2271–95

[15] Sutopo and Waldrip B 2014 Impact Of A Representational Approach On Students'reasoning And Conceptual Understanding In Learning Mechanics *Int. J. Sci. Math. Educ.* **12** 741–65

[16] Susac A, Bubic A, Kazotti E, Planinic M and Palmovic M 2018 Student understanding of graph slope and area under a graph: A comparison of physics and nonphysics students *Phys. Rev. Phys. Educ. Res.* **14** 20109

[17] Knight R D, Jones B and Field S 2009 *College Physics: A Strategic Approach* (Pearson

Education, Inc)

[18] Xu W, Liu Q, Koenig K, Fritchman J, Han J, Pan S and Bao L 2020 Assessment of knowledge integration in student learning of momentum *Phys. Rev. Phys. Educ. Res.* **16** 10130

[19] Paas F and van Merriënboer J J G 2020 Cognitive-Load Theory: Methods to Manage Working Memory Load in the Learning of Complex Tasks *Curr. Dir. Psychol. Sci.* **29** 394–8

Journal of Physics: Conference Series **2126** (2021) 012011 doi:10.1088/1742-6596/2126/1/012011

Development of Thermal Energy Conversion Devices into Electrical Energy as Physics Learning Teaching Aids

D K Umam, S Saehana*, A Kade, and Yunanli

[1]Sub-Department of Physics Education, Faculty of Teacher Training and Education, Universitas Tadulako, Jl. Soekarno Hatta KM. 9 Kampus Bumi Tadulako Tondo Palu-Central Sulawesi, Indonesia

*sahrulsaehana@gmail.com

Abstract. This research aimed to produce a device for converting heat energy into electrical energy as a teaching aid for physics. This research process is development research, which refers to the Research and Development (R&D) model. The instrument used was a questionnaire with a five Likert scale, which addressed media experts, material experts, physics teachers in SMA Negeri 2 Sigi, and class XII MIA SMA Negeri 2 Sigi students. The product quality data obtained were analyzed using descriptive analysis. Based on the analysis of the media expert's assessment, the quality of the product obtained a percentage value of 87.69% with the interpretation of "Very Good". Based on the analysis of the assessment by material experts, the quality of the product obtained a percentage value of 84.44% with the interpretation "Very Good". For the results of the physics teacher's response, a percentage value of 86.00% was obtained with the interpretation "Strongly Agree". For the results of the responses of class XII MIA students in the limited test, a percentage value of 83.56% was obtained with the interpretation "Strongly Agree". The test results and questionnaire analysis showed that the heat energy conversion device developed into electrical energy was feasible to use.

1. Introduction

Physics learning places concepts and practicum as the center of learning because the Physics material is closely related to abstract concepts, so it is necessary to involve hands-on activities in topic delivery. The demonstration given in the form of practicum, either using simulation or laboratory practicum, is an effort the teacher uses to connect the concepts of Physics with real life and make learning more meaningful [1]. In addition to hands-on activities, teaching aids or learning media are also a part that can deliver abstract concepts to be more concrete and realistic [2]. The use of teaching aids can also increase student engagement because they feel happy and enthusiastic in using tools related to everyday life [3]. Besides being used as equipment to assist teachers in transferring learning materials, teaching aids are also intended to increase interest and motivation in learning and make it easier to understand the concepts being studied [4]. Increasing engagement in learning is essential because it is related to learning outcomes and performance [5,6].

One of the materials that contain abstract concepts but is closely related to everyday life is the concept of electricity [7,8]. Several previous studies have shown the positive influence of virtual and real laboratories on materials about electricity on student learning outcomes [9–11]. However, the majority of schools in Indonesia that have limited facilities cannot apply laboratory practicum to students. Because of these limitations, it is necessary to develop accessible and inexpensive teaching aids.

This study aims to develop teaching aids for the conversion system of heat energy into electrical energy. These systems are commonly referred to as direct energy converters (direct-energy converters). This teaching aid can solve the limitations of the material for electricity generators, which is indispensable in learning physics. It is necessary to enrich students' knowledge about sources of electrical energy other than chemical, nuclear, and solar energy. By understanding the material, students have a more significant opportunity to find alternative sources of electrical energy in their daily lives.

2. Method

This research belongs to the type of research and development known as Research and Development (R & R&D), which is a research model used to produce certain products and test their effectiveness [12]. The research began with the design process or a quick overview of the tool. Then, the preparation of the tools and materials to be used and the making of tool designs. Based on the results of consultation and guidance, the tool's design turned into a complete product.

The tool designed was then validated by two expert validators from Physics Education of FKIP Tadulako University using an assessment instrument in the form of responses that will be used as a reference to improve the product. After the revision stage was carried out, it was continued with a limited trial conducted by physics teachers and 30 students of class XII MIA SMA Negeri 2 in Sigi district by filling out a response scale about three months.

The data analysis technique of the research results went through several stages. First, collect questionnaire data, then identify and group according to the assessment and answers on the questionnaire. Second, the data is processed in the form of a percentage. Third, the data were analyzed using quantitative descriptive techniques expressed in the distribution of scores and percentages against the rating scale categories [1], as shown in Table 1.

Table 1. Lickert scale and interpretation of product evaluation

Percentage (%)	Score Scale	Interpretation
0 – 20	1	Very Bad
21 – 40	2	Deficient
41 – 60	3	Sufficient
61 – 80	4	Good
81 - 100	5	Very Good

3. Results and Discussion

3.1. Observation Results of Electrical Energy

The research results that the researchers developed were a device for converting heat energy into electrical energy as a teaching aid for physics. This tool can generate electrical energy by changing the temperature difference between the two sides (hot and cold sides heatsink). The results obtained when using the tool were that the LED (as an indicator) could light up and see the current (I) and voltage (V) using a multimeter.

The observations of electrical energy produced by heat sources of candles with different amounts can be seen in Tables 2, 3, 4 and 5.

Table 2. Results of observation of electrical energy produced with a heat source of 1 candle

No.	**Tp (^{O}c)**	**Td (^{O}c)**	**Δt (^{O}c)**	**V (Volt)**	**I (A)**	**P (Watt)**
1	52	40	12	2,86	0,00090	0,00257
2	52	40	12	2,86	0,00090	0,00257
3	52	41	11	2,84	0,00088	0,00250
4	51	41	10	2,81	0,00079	0,00222
5	51	41	10	2,80	0,00078	0,00218
Mean Score	51,6	40,6	11	2,834	0,00085	0,00241

Table 3. Observation results of electrical energy produced with 2 candles heat source

No.	Tp (°c)	Td (°c)	Δt (°c)	V (Volt)	I (A)	P (Watt)
1	77	41	36	10,20	0,00362	0,0369
2	77	42	35	10,08	0,00334	0,0337
3	77	42	35	10,04	0,00333	0,0334
4	76	41	35	10,00	0,00332	0,0332
5	74	40	34	9,98	0,00330	0,0329
Mean Score	76,2	41,2	35	10,06	0,00338	0,0340

Table 4. Observation results of electrical energy produced with 3 candles heat source

No.	Tp (°c)	Td (°c)	Δt (°c)	V (Volt)	I (A)	P (Watt)
1	92	41	51	13,61	0,00636	0,0866
2	92	42	50	13,60	000630	0.0860
3	91	42	49	13,60	0,00628	0,0854
4	91	42	49	13,60	0,00626	0,0851
5	90	41	49	13,60	0,00625	0,0850
Mean Score	91,2	41,6	49,6	13,602	0,00629	0,0856

Table 5. Observation results of electrical energy produced with 4 candles heat source

No.	Tp (°c)	Td (°c)	Δt (°c)	V (Volt)	I (A)	P (Watt)
1	99	46	53	13,96	0,01448	0,202
2	99	46	53	13,95	0,01448	0,202
3	99	46	53	13,95	0,01448	0,202
4	98	46	52	13,95	0,01447	0,202
5	98	46	52	13,92	0,01447	0,201
Mean Score	98,6	46	52,6	13,946	0,01447	0,202

The relationship between temperature and the potential difference can be seen in Figure 1. On the other hand, the relationship between temperature difference and electric current is presented in Figure 2.

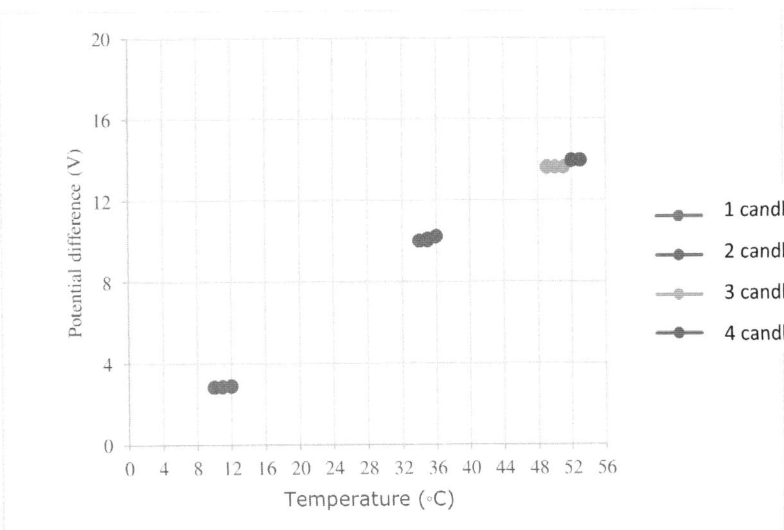

Figure 1. Graph of the relationship between temperature and potential difference

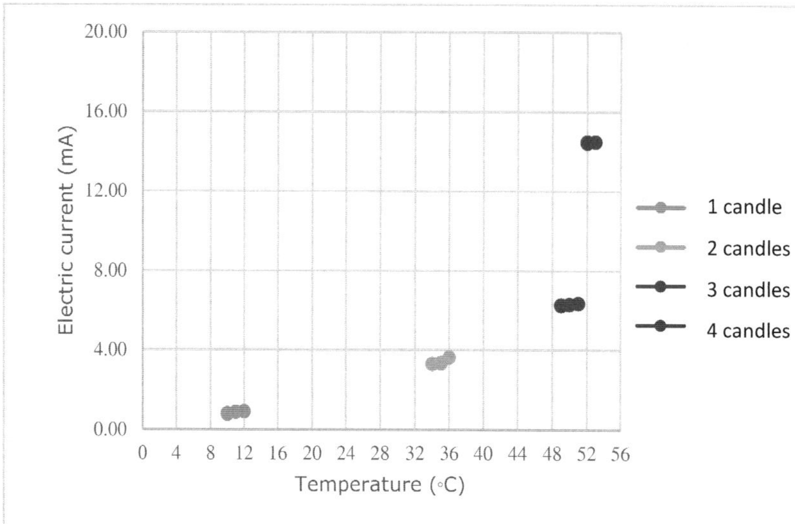

Figure 2. Graph of the relationship between temperature difference and electric current

After revising, the development of the tool was then continued, with validation from media experts and material experts. The products produced by researchers in the tool's design for converting heat energy into electrical energy are shown in Figure 3.

Figure 3. Thermal energy conversion device into electrical energy

3.2. Assessment Results of Thermal Energy Conversion Devices into Electrical Energy

3.2.1 Assessment by media experts

The validation results by media experts using several aspects can be seen in Table 6. The results of the assessment on each aspect were as follows: the efficiency of teaching aids was in a "very good" category (VG) with a percentage score of 90%; aesthetic aspects gained "good" criteria (G) with a percentage score of 80%; the durability aspect of the teaching aids was in "good" criteria (G) with a percentage score of 80%; the safety aspect for students included in "very good" criteria (VG) with a percentage score of 85%; and the mean score of the four aspects was 87.69% with "very good" criteria (VG).

Table 6. Assessment Results by Media Experts

No.	Aspect	Criteria	Result
1	Efficiency	Ease of Assembling	Good
		Ease of Use	Good
		Ease of Transfer	Very Good
		Ease of Saving	Very Good

2	Aesthetics	Shape	Good
		The Suitability of Teaching Aids With Students' Physical Size	Good
3	Props Durability	Not Easily Loose, Broken or Destroyed When Used	Good
		Installation Accuracy of Each Component	Good
		Component Resistance in its Original Position.	Good
4	Safety for Students	Made of Safe Materials (Not Sharp)	Good
		Safe Construction for Students (Not Easy to Collapse)	Very Good
		The use of tools does not require special treatment (wearing a mask or gloves).	Very Good
		No Effect of Harmful Chemicals (Non-Flammable, Non-Irritating).	Very Good

3.2.2. Assessment by material experts

The results of the material expert assessment carried out using several aspects can be seen in Table 7.

Table 7. Assessment Results by Material Experts

No.	Aspect	Criteria	Result
1	The Relationship between Teaching Aids and Teaching Materials	Show Physical Phenomena	Good
		Required in Learning Renewable Energy Source Materials	Good
		Generate Concepts from Props	Good
		Explain the concept of Renewable Energy	Good
2	Educational Value	The Suitability of Teaching Aids With the Intellectual Development of Students	Very Good
		Easy for Students to Understand Concepts	Good
		Teaching Aids Suits the Needs in Learning	Good
		Ease For Learners To Improve Skills (Thinking, Speaking And Moving)	Very Good
		Fostering Motivation for Students to Study More Actively	Good

The results of the assessment of each aspect can be seen in Figure 4. Overall, all aspects were in the 80 percent or included in the "Very Good" category. The assessment of the relationship between teaching aids and teaching materials was in the "Very Good" (VG) category with a percentage score of 80%. Aspects of educational value got "Very Good" (VG) criteria with a percentage score of 88%.

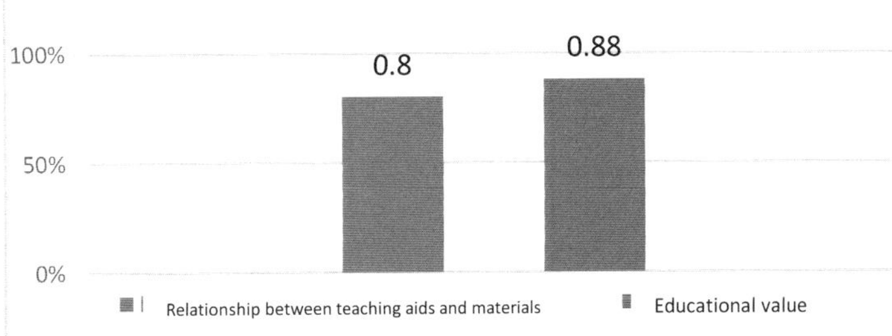

Figure 4. Material experts' assessment chart

3.2.3. Assessment by physics teacher

The results of the physics teacher's assessment of the developed product can be seen in Table 8. The results of the assessment of each aspect were as follows: the assessment of the relationship between teaching aids and teaching materials was in the "Very Agree" (VA) category with a percentage score of 85%; the assessment of the educational value got the category of "Agree" (A) with a percentage score of 72%; assessment of the efficiency of teaching aids, and safety aspects got a perfect score of 100% with the category of "Very Agree" (VA); while the aesthetic and durability aspects included the "Agree" criteria (A) with a percentage score of 80%; and the percentage mean score of the six aspects was 86.36% with the criteria of "Very Agree" (VA).

Table 8. Results of assessment by physics teacher

No.	Aspect	Criteria	Result
1	The Relationship between Teaching Aids and Teaching Materials	Show Physical Phenomena	Very Agree
		Required in Learning Renewable Energy Source Materials	Agree
		Generate Concepts from Props	Agree
		Explain the concept of Renewable Energy	Agree
2	Educational Value	The Suitability of Teaching Aids With the Intellectual Development of Students	Agree
		Easy for Students to Understand Concepts	Very Agree
		Teaching Aids Suits the Needs in Learning	Uncertain
		Ease For Learners To Improve Skills (Thinking, Speaking And Moving)	Uncertain
		Fostering Motivation for Students to Study More Actively	Uncertain
3	Efficiency	Ease of Assembling	Very Agree
		Ease of Use	Very Agree
		Ease of Transfer	Very Agree
		Ease of Saving	Very Agree
4	Aesthetics	Shape	Agree
		The Suitability of Teaching Aids With Students' Physical Size	Agree
5	Props Durability	Not Easily Loose, Broken or Destroyed When Used	Agree
		Installation Accuracy of Each Component	Agree
		Component Resistance in its Original Position.	Agree
6	Safety for Students	Made of Safe Materials (Not Sharp)	Very Agree
		Safe Construction for Students (Not Easy to Collapse)	Very Agree
		The use of tools does not require special treatment (wearing a mask or gloves).	Very Agree
		No Effect of Harmful Chemicals (Non-Flammable, Non-Irritating).	Very Agree

3.2.4. Assessment by students

Aspects assessed by class XII MIA students included learning motivation and understanding of the concept of renewable energy sources using teaching aids, the operation and performance of teaching aids, and the quality of teaching aids. The results of the assessment by students are presented in Table 9. From the data of the questionnaire analysis, it can be seen that the average students' responses were "Very agree" with a percentage above 80% for all aspects assessed. In other words, class XII MIA students of SMA Negeri 2 Sigi responded very well to the existence of a heat energy conversion device into electrical energy as a teaching aid for physics learning designed by researchers.

Table 9. Assessment Results by Class XII MIA Students

No.	Criteria	Response Criteria	Result
1	Motivation to learn and understand the concept of renewable energy sources with teaching aids	It is easier for me to understand the concept of physics in renewable energy sources with the presence of teaching aids.	Very agree
		My enthusiasm for learning Physics has increased with the use of teaching aids as a medium of learning.	Very agree
		I enjoy learning Physics with props.	Very agree
		I am more active in learning with teaching aids.	Very agree
		I collaborate with friends when learning using props	Very agree
		The teaching aids can increase my knowledge about the application of physics.	Very agree
		The existence of teaching aids in learning made me interested in trying and making teaching aids as exemplified.	Agree
2	Operation and performance of props	The props are easy to operate.	Agree
		Props work well when used.	Agree
		Props are suitable for use for class XII students	Agree
3	Props quality	The teaching aids helped me to gain an understanding of the concept of renewable energy sources.	Very agree
		Teaching aids can present the concept of renewable energy sources.	Very agree

Based on the observation results of heat energy converted into electrical energy using a device with a thermoelectric cooler 12706 as the base material, it follows the theory that the greater the temperature difference between the two sides, the greater the electrical energy produced. From the observations on four experiments using 1, 2, 3, and 4 candles as a source of heat energy, the amount of heat energy affects the amount of electrical energy produced. From the graph of the observations, it can be seen that the voltage generated by the heat energy conversion device into electrical energy is proportional to the temperature difference between the two junctions. The results obtained that the average potential difference in degrees Celsius was 0,272 V/^{0}C. While the results obtained for the average electric current per degree Celsius was 1,44 x 10^{-4} A/^{0}C, and the amount of electric power obtained per degree Celsius was 1,68 x 10^{-3} Watt/°C.

4. Conclusion

Based on the results of data analysis and discussion, it can be concluded that the device for converting heat energy into electrical energy as a teaching aid for physics meets the Very Good criteria and is suitable for use. The produced energy converter specifies the relationship between the temperature difference of the two sides and the increase in potential difference and electric current. The average value for the potential difference per degree Celsius was equal to 0,272 V/^{0}C, the electric current per degree Celsius was 1,44 x 10^{-4} A/^{0}C, and the magnitude of the electric power per degree Celsius was 1,68 x 10^{-3} Watt/^{0}C.

Acknowledgement
We highly acknowledge the support of the Program Kompetisi Kampus Merdeka (PK-KM) Grant of the Education, Culture, Research, and Technology of the Ministry of the Republic of Indonesia, with funding identity : 119/E1/KM.05.03/2021.

References

[1] Saehana S, Ali M and Supriyatman S 2019 T *J. Pendidik. IPA Indones.* **8** 241–246
[2] Franklin S and Peat M 2005 *Int. J. Contin. Eng. Educ. Life Long Learn.* **15** 67–78
[3] Hani U S 2017 Pengembangan media pembelajaran sebagai alat peraga penerapan konsep hukum pascal untuk peserta didik kelas viii smp
[4] Lestari L P 2006 *Keefektifan pembelajaran dengan penggunaan alat peraga dan lembar kerja siswa (lks) terhadap hasil belajar matematika dalam pokok bahasan bangun segiempat pada siswa kelas vii semester 2 di smp muhammadiyah margasari kabupaten tegal tahun pelajaran 2005/2* (Universitas Negeri Semarang)
[5] Lee J S 2014 *J. Educ. Res.* **107** 177–185
[6] Zacharia Z C 2007 *J. Comput. Assist. Learn.* **23** 120–132
[7] Izzah A and Saehana S 2021 *J. Phys. Conf. Ser.* **1760**
[8] Rohman A, Komang Werdhiana I and Saehana S 2021 *J. Phys. Conf. Ser.* **1760**
[9] Ajredini F, Izairi N and Zajkov O 2017 *Eur. J. Phys. Educ.* **5** 59–70
[10] Baser M and Durmus S 2010 *Eurasia J. Math. Sci. Technol. Educ.* **6** 47–61
[11] Farrokhnia M R and Esmailpour A 2010 *Procedia-Social Behav. Sci.* **2** 5474–5482
[12] Sugiyono 2012 *Metode Penelitian Pendidikan* (Bandung: Alfabeta)

Development of simple science kits as dynamic electricity learning media for junior high school

S Saehana[1*], I K Werdhiana[1], N Tuljannah[1], and A Izzah[1]

[1]Sub-Department of Physics Education, of Faculty Teacher Training and Education, Universitas Tadulako, Jl. Soekarno Hatta KM. 9 Kampus Bumi Tadulako Tondo Palu-Central Sulawesi, Indonesia.

*Corresponding author: sahrulsaehana@gmail.com

Abstract. The aim of this study is to produce Simple IPA KIT as science learning media for dynamic electricity material. This Simple IPA KIT contains several puzzle blocks that could be set into several electrical circuits. The type of the study was Research and Development, which refers to the model of Puslitjaknov (a team in the Center for Policy and Inovation Studies). The five-scales Likert questionnaire were used to obtain the data. The product was assessed by the media and material experts, physics teachers and high school students. The result of media experts showed a percentage value of 85%, in "Very Good" category, while it was 80% in "Good" category for material expert assessment. Meanwhile, the responses of teachers and students showed 95.6% and 84%, respectively, with "Very Good" category. It indicates that the Simple KIT media is suitable for learning media.

1. Introduction

The abstract concepts in dynamic electrical topics make them challenging to be fully understood by students [1–3]. Moreover, teachers tend to use teacher-centered methods and textbooks in learning processes that drive students unmotivated [4]. These methods are ineffective, especially when teaching about physics concepts that are abstract concepts. For instance, when the researchers did observations at the preliminary research activities in some public schools in Central Sulawesi, teachers explained the concept of electricity as a part of dynamic electricity topic, while students only listened to the description of the definition of electricity, the electrical propagation, and electrical circuit.

IPA KIT-Media is an integrated laboratory instrument box for science experiments [5]. The integration of kits in science education does not only benefit the students [6,7], but also advantages the teachers [8]. It enables the teachers to create innovate and active learning environment because the kits allows the teachers to conduct a variety of learning activities [9]. The role of kits in science learning does not only help students understand the concepts, but also demonstrate how the concepts are applied through laboratory activities [10]. A study of Jones et al [11] found that students who learned consistently using kits in science learning are more likely to have scientific skills, such as designing and implementing laboratory investigations, results presentation, and data analysis. The use of kits, specifically in physics learning can help students in a complex and challenging concept, such as electricity [12]. It improve students' enthusiasm and motivation in learning because they consider learning as playing using the kits [13]. It was supported by a study of Miller [14] who found that children preferred to more creative and entertaining apps rather than pedagogical-based apps.

IPA KIT-Media for junior high school consists of 4 packages, namely KIT for Mechanics, KIT for Heat and Hydrostatic, KIT for Optics, and KIT for Magnetic Electricity. However, the integration of IPA KIT-Media in science learning is not very common. It is unfortunate since Simple KIT-based learning media was beneficial for students in learning dynamic electricity [8]. Teachers who were included in our preliminary activities, indicated that the use of IPA KIT-Media has not been optimal due to several reasons. Firstly, there is not enough equipment for certain electrical concepts or some components of an experiment are missing. Secondly, teachers admitted that they lack the ability to operate the tools. Thirdly, the KIT is quite expensive thus the teachers only could expect that the school will provide it. To solve these problems, researchers intended to provide innovation of the Simple IPA KIT that is affordable yet still beneficial for science learning.

If most research and development (R&D) studies in electricity concepts used only electrical components in their product (i.e [6, 9]), the developed kits in this study used puzzle blocks as the main material. Each of the puzzle blocks was designed by gluing a sheet of zinc on it and it serves as connectors that deliver currents from the source to the indicator to be measured. Students could use these puzzle blocks to set different types of electrical circuits such as Ohm's Law Circuits, Series Electrical Circuits, Parallel Electrical Circuits, and Kirchoff's Law I circuits. Cardozo et al who conducted a study about the effect of puzzles in the learning process revealed that the activity with the puzzle assisted students to understand the topics better [10]. This research was also inspired by the results of a study of Stezik et al who found that students preferred to learn physics with the addition of puzzle games rather than lecture-only methods [11].

Accordingly, the objective of this research was to produce the Simple IPA KIT using the puzzle blocks. The product was expected to fulfil the limitation of science laboratory equipment in high schools, and eventually could improve students' knowledge about dynamic electricity and their skills arranging electrical circuits [15]. The complete Simple KIT IPA contains several resistors, a switch, a battery, a simple indicator lamp (mini incandescent lamp), puzzle blocks, multimeters, and the guidebook in a box thus it is portable.

2. Methods

2.1 Type of Research

It was a Research and Development (R & D) study. It employed the procedures developed by Puslitjaknov [16] (the team in the Center for Policy and Inovation Studies) whose the stages were more efficient]. This model belongs to the Borg & Gall and simplified by Puslitjaknov into five stages, namely analysis, early development product, expert validation and revision, limited field testing and revision, and big-scale field testing. However, this research only applied four stages, which excluded the big-scale field testing due to the limited time and cost. As the details of the research scheme can be observed in Figure 1.

2.2 Location and Subject of The Research

This research took place in SMP Negeri 19 Palu, Central Sulawesi for the limited field testing stage. The production process required a duration of 6 months. All the stage processes included one each of media and material experts, and ten physics teachers and ten ninth graders to assess the feasibility of the developed product.

2.3 Data Collection Techniques

Data were collected through interview and questionnaire distribution. Interview was conducted with the physics teacher to acquire the actual problems. Meanwhile, the questionnaire was used to assess the product from the perspective of media and material experts, teachers and students. The questionnaire was a 5-point Likert scale. Likert scale as a scaling technique is widely used primarily to measure attitudes, opinions or perceptions about themselves or their group of people who are related to something [17]. This scale is often referred to as a summated scale which contains a number of statements with response categories. The data collected were analyzed into percentage [18], and converted into qualitative data based on categories in Table 1 [19].

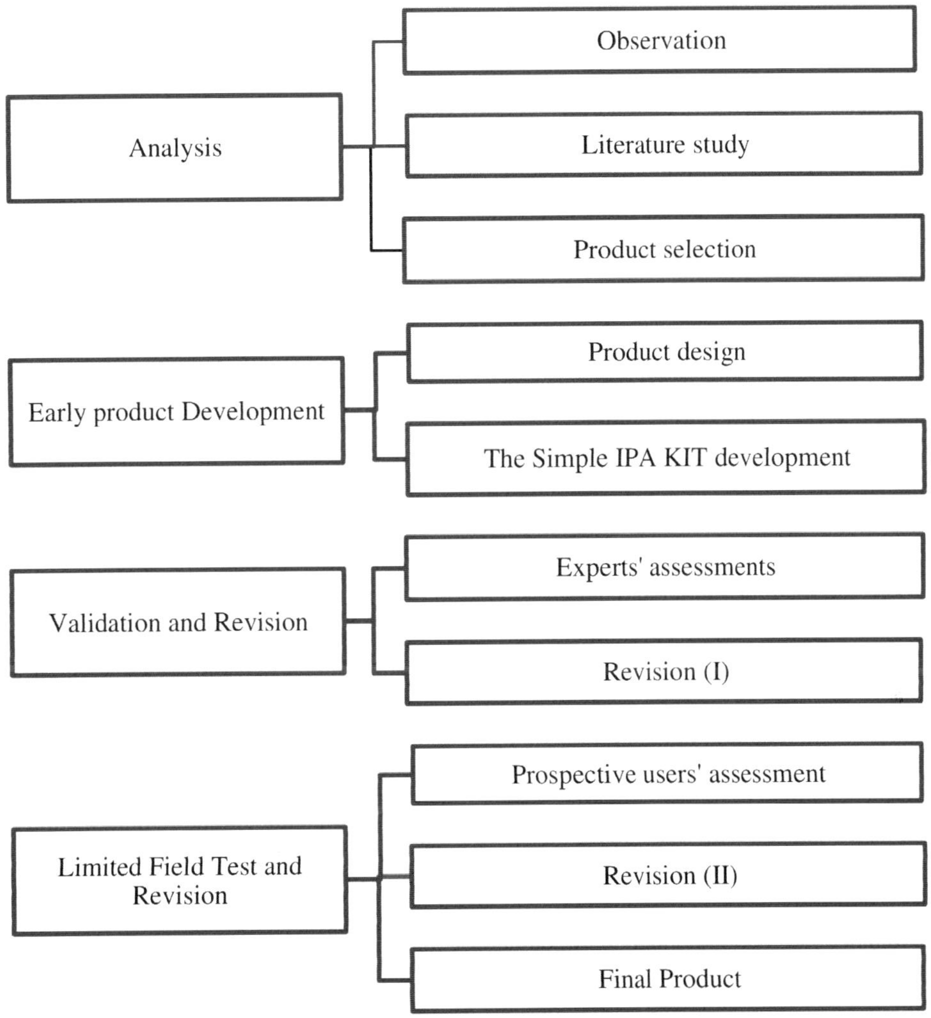

Figure 1. The Simple IPA KIT development scheme

Table 1. Likert Scale Category

Percentage	Scale	Category
0% - 20%	1	Very Poor
21% - 40%	2	Poor
41% - 60%	3	Enough
61% - 80%	4	Good
81% - 100%	5	Very Good

Journal of Physics: Conference Series **2126** (2021) 012012 doi:10.1088/1742-6596/2126/1/012012

3. Result and Discussion

In developing simple IPA KIT using puzzle blocks, researchers first create an initial design or sketch of the electrical circuit (see Figure 2a). Its working principle uses two 9V batteries as a voltage source, and the puzzle blocks as materials that allowed electric currents from the source to the multimeters. To make puzzle blocks work as connectors, researchers laid zinc sheets on them and then strengthened by wires. After that, the electrical circuit components such as resistors, switches, and socket cables were installed on puzzle blocks. The following Figure 2 represents the initial designs and examples of usable puzzle blocks.

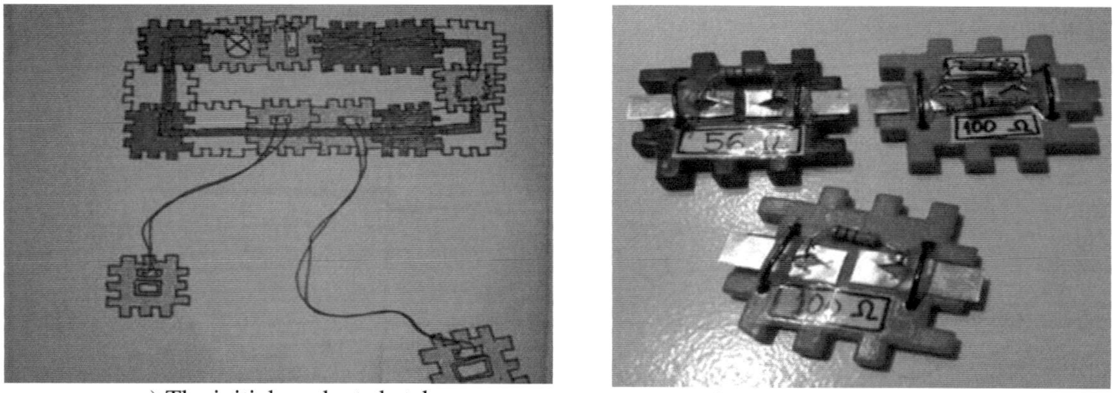

a) The initial product sketch b) Puzzle blocks with resistors

Figure 2. The initial designs of the product and the examples of usable puzzle blocks

Puzzle blocks that have been developed as presented in Figure 2b, were arranged into 4 types of electrical circuits such as Ohm's Law Circuits, Series Electrical Circuits, Parallel Electrical Circuits, Combination Series-Parallel Circuits and Kirchoff's Law I circuits. The appearance of Simple IPA KIT for each circuit can be seen in Figure 3-6.

3.1 Ohm's Law Circuits

This display was the process of examining the voltage on the resistor component of 100 ohms and the electric current.

Figure 3. Puzzle blocks arrangement for Ohm's Law circuit

3.2 Series and Parallel Electrical Circuit

This process determined the current and voltage used a resistor of 56 ohm and 100 ohm for series circuits as well as parallel circuits.

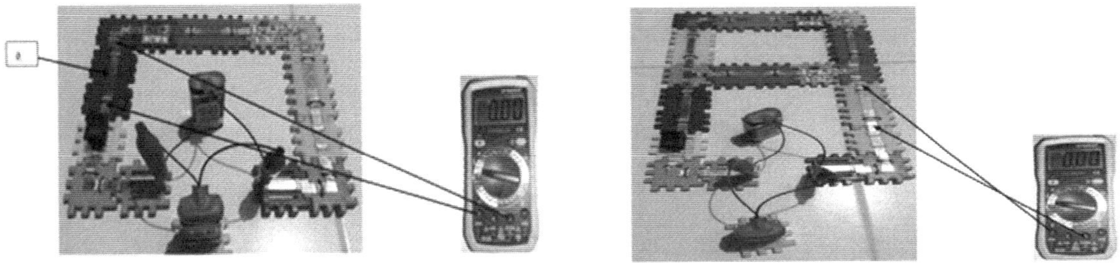

a) Series circuits b) Parallel circuits

Figure 4. Puzzle blocks arrangement for series circuits as well as parallel circuits.

3.3 The Combination of Series and Parallel Electrical Circuits

The determination of current and voltage in this circuit was to parallel a resistor 47 ohm and 100 ohm then arranged in series with a resistor of 56 ohms.

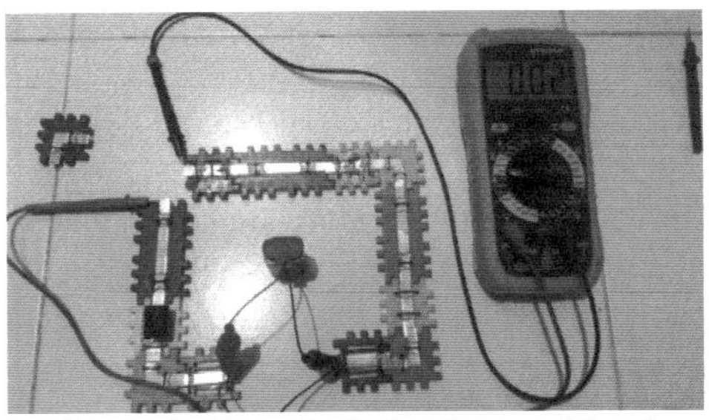

Figure 5. Electrical circuits combination

3.4 Kirchoff's Law I Circuits

This display was an electrical circuit that aims to find out the amount of electric current that enters and exits.

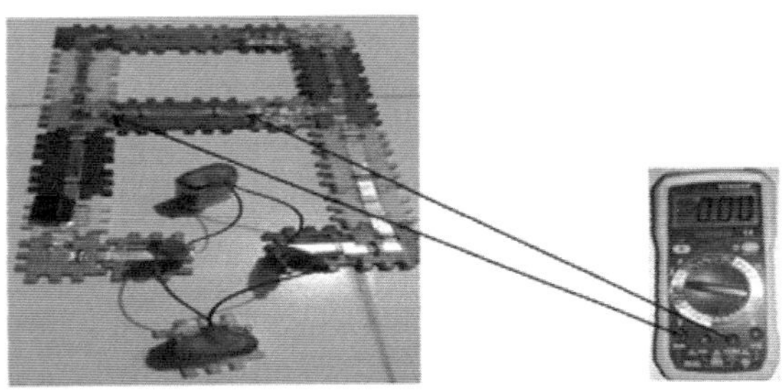

Figure 6. Kirchhoff's Law I circuits

Moreover, the guidebook was also provided. It consists of some parts such as instructions to measure the current (I) and the voltage (V), relevant information, components of practicum puzzles, experimental purposes, data processing, final questions and conclusions. All tools were further validated and revised by media and material experts. The responses were used as references to improve the quality of the product. The results of the validation tests are displayed on Figure 7.

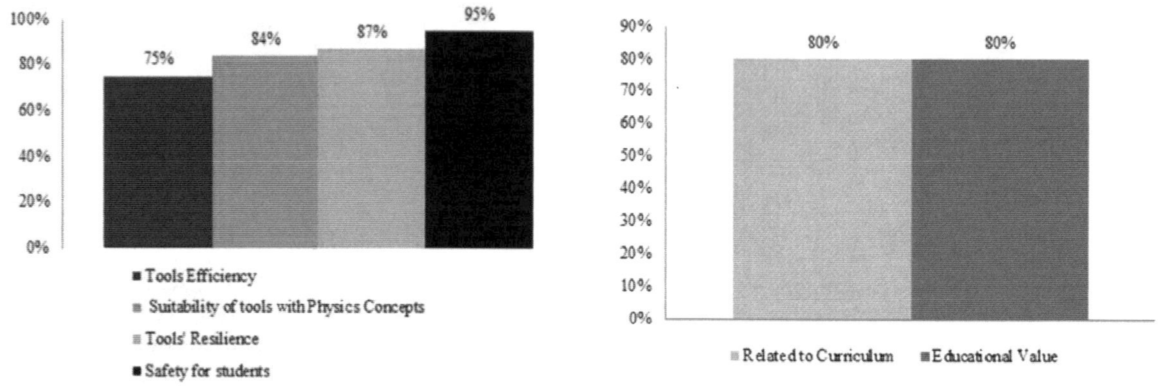

a) Results from the media expert b) Result from the material expert

Figure 7. Assessment results from media and material validators

Figure 7a shows that the average value of the overall aspects for media validation was in the "very good" category with the percentage of 85%. Meanwhile, the assessment for the contents obtained a "good" category with the percentage of 80%. These values show that The Simple IPA KIT was feasible to undergo field-testing. Revision was made in the puzzle blocks, including making the components permanently attached to the blocks, and giving clear instructions in the guidebook and worksheets for each experiment. According to Puslitjaknov, product trials are significant in research development. Product trials aim to find out the effectiveness of the product. In the limited field-test, data were obtained from physics teachers and students as the prospective product users. The results of the assessment can be seen in Figure 8.

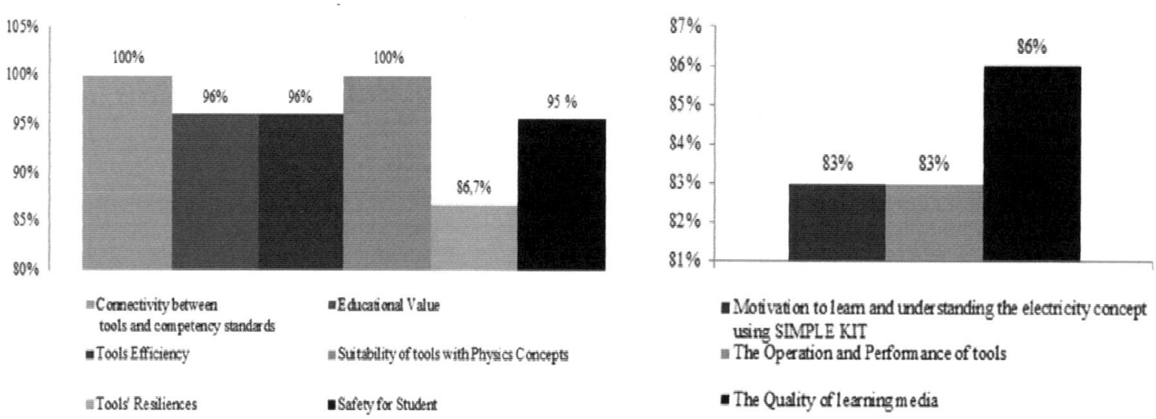

a) Results from teachers' assessment b) Result from students' assessment

Figure 8. Limited field-test results

Both the results of teachers and students' responses were in the "very good" category, with the percentage of 95.6% and 84% respectively. There was no additional revision in this limited test. Students were very excited with the developed product. In the first trial of Ohm's Law, students needed about three minutes and 43 seconds to arrange the circuit. Meanwhile, they took 3.84 minutes for the series circuits, 05.27 minutes the parallel circuits, 3.40 minutes for Kirchoff's Law I, and 2.56 minutes for the series-parallel circuits with the help of a guidebook only.

The final product of this research was similar to the product produced by Dwiokta and Desnita who developed a data logger-based practicum using almost the same electrical circuits [20]. However, their final product uses many puzzles and must be set separately, while this study only used a few puzzles that can be disassembled and rearranged according to the needs in the experiment. Another benefit of the final product is its simple packaging. All components are fitted in a box including multimeters. The product is smaller in size and lighter thus it is very portable. It allows it to be carried anywhere thus the students even can do the hands-on activities outside the schools.

4. Conclusions

The simple IPA KIT as a science learning media consisting of circuit puzzles, multimeters, and guidebook in a toolbox has been developed. The assessments of the media and material experts as the validators, and physics teachers and students as the prospective users, show that the final product has met the criteria of feasibility of learning media. As a result, the product is feasible to be used in learning activities. However, the developed device still has one weakness which is the puzzle interblock connectors that were not very tight resulting in the instability when measuring the voltage (V) and current (I) in the multimeter. Further studies are also required to determine the effectiveness of this product to students' learning outcome.

Acknowledgement

We thank kindly Hibah Penelitian Terapan Kompetitif Nasional DIKTI and Program Kompetisi Kampus Merdeka (PK-KM) of Ministry of Education, Culture, Research and Technology of The Republic of Indonesia, under the funding 302/E4.1/AK.04/PT/2021 and 119/E1/KM.05.03/2021, respectively.

References

[1] Ekici E 2016 "Why Do I Slog Through the Physics?": Understanding high school students' difficulties in learning physics *J. Educ. Pract.* **7** 95–107
[2] Reddy M V B and Panacharoensawad B 2017 Students Problem-Solving Difficulties and Implications in Physics : An Empirical Study on Influencing Factors *J. Educ. Pract.* **8** 59–62
[3] Halim A, Lestari D and Mustafa 2019 Identification of the causes of misconception on the concept of dynamic electricity *Journal of Physics: Conference Series* vol 1280
[4] Sihaloho R R, Sahyar S and Ginting E M 2017 The Effect of Problem Based Learning (PBL) Model toward Student's Creative Thinking and Problem Solving Ability in Senior High School *IOSR J. Res. Method Educ.* **7** 11–8
[5] Rahayu S, Harjono A and Gunada I W 2019 Pelatihan Penggunaan KIT IPA Bagi Guru dan Siswa SMP N 1 Sakra Lotim *J. Pengabdi. Masy. Sains Indones.* **1**
[6] Budiastra A A K, Kusmawan U, Wicaksono I and Kartimi 2020 The Use of Natural Sciences Kits in Distance Learning for Higher Education of Bachelor of Elementary School Teacher Education Program *Adv. Soc. Sci. Res. J.* **7** 54–69
[7] Gumala Y, Nurkaeti N, Aryanto S, Syaodih E and Samsudin A 2020 The Influence Of Using Kit Of Science For Kids To Elementary School Students' Concept Mastery *PrimaryEdu - J. Prim. Educ.* **4** 74–82
[8] Sherman A and MacDonald A 2008 The Use of Science Kits in the Professional Development of Rural Elementary School Teachers. *Sci. Educ. Rev.* **7** 91–105
[9] Jaimini N 2014 Science Kits as Resource: Some Pedagogical Considerations *J. Res. Method Educ.* **4** 16–9
[10] Suryani N A, Sakti I and Purwanto A 2018 Perbedaan Hasil Belajar Antara Model Pembelajaran Clis (Children's Learning In Science) Dengan Menggunakan Media Kit IPA di SMP Negeri 21 Kota Bengkulu *PENDIPA J. Sci. Educ.* **2** 113–6
[11] Jones G, Robertson L, Gardner G E, Dotger S and Blanchard M R 2012 Differential use of elementary science kits *Int. J. Sci. Educ.* **34** 2371–91
[12] Rogosic R, Heidt B, Passariello-Jansen J, Bjornor S, Bonni S, Dimech D, Arreguin-Campos R, Lowdon J, Jimenez Monroy K L, Caldara M, Eersels K, Van Grinsven B, Cleij T J and Dilien H 2021 Modular Science Kit as a support platform for STEM learning in primary and secondary

school *J. Chem. Educ.* **98** 439–444

[13] Peppler K, Wohlwend K, Thompson N, Tan V and Thomas A M 2019 Squishing Circuits: Circuitry Learning with Electronics and Playdough in Early Childhood *J. Sci. Educ. Technol.* **28** 118–32

[14] Miller T 2018 Developing numeracy skills using interactive technology in a play-based learning environment *Int. J. STEM Educ.* **5** 1–11

[15] Izzah A and Saehana S 2021 Developing video-based learning for the application of ohm's law towards conductive transparent layer *Journal of Physics: Conference Series* vol 1760 pp 1–9

[16] Puslitjaknov T 2008 *Metode Penelitian Pengembangan*

[17] Silalahi U 2006 *Metode Penelitian Sosial* (Bandung: Unpar press)

[18] Arikunto S 2011 *Prosedur Penelitian Pendekatan Suatu Praktik* (Jakarta, Indonesia: Rineka Cipta)

[19] Widoyoko E P 2016 *Evaluasi Program Pembelajaran)* (Yogyakarta: Pustaka Pelajar)

[20] Karlina D M P, Desnita and Nasbey H 2015 Pengembangan Set Praktikum Berbasis Data Logger Pada Pembelajaran Fisika Sma Kelas XII Materi Rangkaian Arus Listrik Searah *Seminar Nasional Fisika (E-Journal) SNF2015* (Universitas Negeri Jakarta) pp 107–12

The influence of the online PhET simulation-assisted using direct instruction on student's conceptual understanding of parabolic motion

P D Lestari[1*] and J Mansyur[1]

[1]Sub-Department of Physics Education, Faculty of Teacher Training and Education Universitas Tadulako, Jl. Soekarno Hatta KM. 9 Kampus Bumi Tadulako Tondo Palu-Central Sulawesi, Indonesia.

*Corresponding author: putridianlestari99@gmail.com

Abstract. This study aimed to investigate the effect of the online PhET simulation-assisted using the direct instruction on students' conceptual understanding of the parabolic motion. It was quasi-experimental research with pretest-posttest control group design. The research subjects were 10th grade students, consisting of 20 students for each experimental and control groups. The data were collected using a concept understanding test. The research data were analyzed quantitatively by using descriptive and inferential statistics. The results show that the average score of the experimental class was higher than the control class. The hypothesis was tested using Mann-Whitney (U-Test) with a significance level of = 0.05. Based on data analysis, it can be concluded that there was a significant effect of online PhET simulation assisted learning model on students' concept understanding on parabolic motion.

1. Introduction

During the pandemic era, the use of technology plays a crucial role in the realm of education. Its utilization, especially in online classes, requires creativity and instructional considerations [1]. Technology application that has been widely used to support the learning process is simulations [2]. One of the interactive simulation platforms that is frequently used in physics learning is Physics Education Technology (PhET) [3]. PhET was developed by the University of Colorado at Boulder, United States to provide various simulations to explain concepts and theories in interactive and interesting ways that can be run from websites using standard web browsers [4][5]. Moreover, the users are also able to download and install the website and use it for offline learning [6].

Virtual laboratory-based is effective in physics learning for both the teachers and students [7][8]. Its application can improve students' interest in physics [9][10], which is heavily correlated with conceptual understanding [11]. Using simulation in learning activities can provide learning experiences while playing [12]. The simulation helps students examine a physical phenomenon or event through an interesting illustration [13][14][15]. Further, the previous study indicated that the use of PhET simulation in direct instruction can increase students' creativity [16].

Direct instruction is a teacher center learning [6]. This model is suggested for learning during the pandemic situation in Indonesia since students are not familiar with the self-study type of learning from their respective houses [17]. Accordingly, teachers' roles in learning become dominant in creating interesting learning activities for students [6]. Teachers should not solely focus only on the topic explanations, but also on the knowledge and skill demonstration thus teachers are required to use a variety of learning media [18].

Students' learning experiences in the direct learning model train students to think, apply the skills they have just acquired, and build their own understanding so that the learning process becomes active [18]. There are five important stages in direct learning, namely the orientation, presentation or demonstration, structured exercises, guided exercises, and independent exercises. The application of direct instruction can encourage students to think and apply the acquired knowledge [18]. In addition, it also can increase students' learning experiences, foster scientific attitudes and teamwork [19]. However, due to the unprecedented situation in the couple years, learning activities have changed from face-to-face to totally virtual learning. Despite the challenging condition, learning processes can still be conducted because of the technology [20]. Some benefits of online learning with the utilization of effective multimedia are improving the quality and the range of quality of education and training, and reducing the cost of education and training [21]. Since the learning activity is conducted through a web network, it can accommodate many participants from different locations and expertise [22].

Students who have concept understanding on a taught topic have the ability to explain it in their words and to apply it in a real-life situation [23]. Conceptual understanding in physics is crucial due to the intercorrelation among the concepts and theories. Students who do not understand a concept will encounter difficulty to understand the next material [24]. One of the physics topics that is challenging for students is parabolic motion [25]. It is complex because students require explanations in the form of graphic images, vectors and real examples in everyday life [26]. Some misconceptions about parabolic motion were about the complex nature of motion. Many students still have misconceptions on the differences between the main principles of kinematics, including position, velocity, acceleration and time [27]. In addition, the use of graphic interpretation, representation, equations and also combining two-dimensional motion in this topic makes it a difficult topic [28]. Students have particular difficulty in analyzing the components of velocity and acceleration on the x-axis and y-axis and making two graphs of the relationship between velocity and time or v-t graphs [29][30]. It was frequently found on students' mistakes during problem solving [31].

Previous researchers mostly conducted direct learning using PhET simulation in face-to-face mode learning. It is rare to find a study that has investigated the impact of PhET simulation integrated in direct instruction virtually. On that account, this study aims to measure the effect of a direct learning model assisted by PhET simulations in virtual learning on students' conceptual understanding in parabolic motion.

2. Methods

This research was a quasi-experiment with pretest-posttest control group design. It was conducted in one of the public schools in the Sigi Regency, Central Sulawesi. A total of 40 students were involved in this study, with 20 students for each experimental and control group. The samples were chosen using simple random sampling by lottery. Students in the experimental class were taught using a direct learning model assisted by online PhET simulation while the control class was only taught using the direct instruction. Both groups went through the same stages of learning, including the material delivering, peer discussion, and presenting and feedback session stage. The difference was on the method used on the material delivering stage. In the control group the teacher provided material in the form of written resources. By doing so, students were able to interpret and provide conclusions from the material they had read. Meanwhile, in the experimental group the teacher guided the students to experiment with parabolic motion using PhET simulation after the material explanation.

Data were collected using a concept understanding test. It was an essay test consisting of 5 questions. The test was used to measure students' understanding of physics concepts both before and after the implementation of the online PhET simulation-assisted using direct instruction. Each question represented one indicator, namely translation indicator for question number 1 and 3, interpretation of question for question number 2, and extrapolation for numbers 4 and 5. These indicators were adopted from a study of Wuryanti et al [23]. The data were analyzed quantitatively using descriptive and inferential statistics. Descriptive statistics were used to describe the achievement of each variable, while inferential statistics were for the hypothesis test [32]. The alternative hypothesis of this study was that the use of direct instruction with PhET simulation virtually has a significant effect on students' conceptual understanding in parabolic motion.

3. Results and Discussion

Based on the descriptive data, there was a difference in conceptual understanding between students in the control and the experimental groups. The pretest results showed that the control class got a higher average score than the experimental class. Meanwhile, the average score of the posttest indicated that the experiment outperformed the control class. Detailed data were presented in Table 1.

Table 1. Pretest-posttest score of the control and experimental group

Description	*Pretest*		*Posttest*	
	Control	Experimental	Control	Experimental
Sample (n)	20	20	20	20
Minimum value	45	40	45	50
Maximum value	80	80	85	95
Average	60	55	64	76

The results of indicator analysis for each question between both groups can be seen in Figure 1.

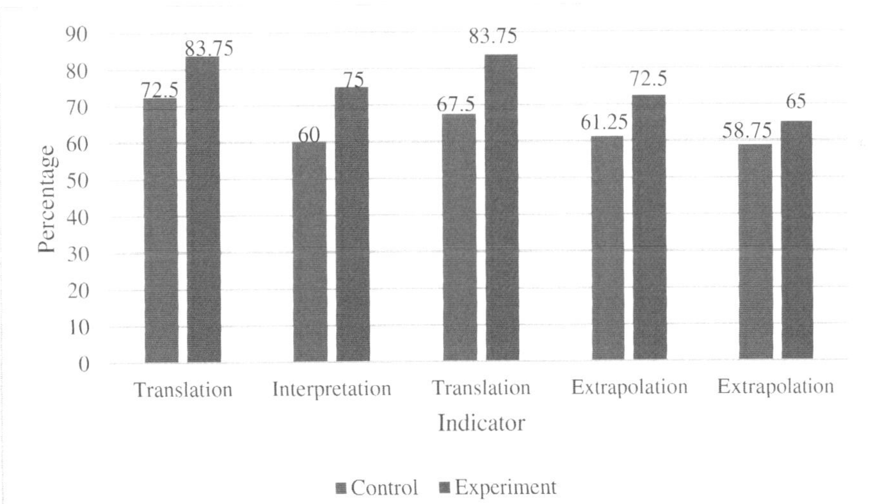

Figure 1. Percentage of indicators of conceptual understanding

Data presented in Table 1 and Figure 1 show that there are differences between the control and treatment group. It is owing to the phases of learning activities that encouraged students to be more active and involved in learning although all the activities were constructed online. The demonstration using the PhET simulation allowed to reduce the level of variance in the study of understanding and to develop accurate conceptual understanding [33]. Moreover, online PhET simulation can improve students' thinking skills [4]. Students in the treatment group outdid the control group because the use of simulation helped them to stay focused on the topic delivering process [34]. It is evident that PhET simulation gives a positive effect in learning [35], in either offline or online mode of learning. Without the use of PhET simulation, students in the control group lacked interest in learning. However, since direct instruction consists of a discussion after the stage of delivering material lessons, it was able to increase students' interest. A group discussion was also used to support the implementation of the learning process [36]. The posttest score of the control group also improved compared to the pretest. It shows that the direct instruction also increased students' learning outcome, although as not as effective as when it was assisted by PhET simulation.

Understanding the concept of parabolic motion in the indicator of translation is quite good. It is shown by the score for the translation aspect which has the highest score compared to the other indicators. Students could describe graphs and pictures verbally correctly. It was in accordance with the study of Jelatu and Kurnila [36] who found that most of their research subjects were in the medium category which meant that they had the ability to translate from verbal to visual forms yet had not done perfectly. Further, the finding was in contrast with the result of Usman and Hussaini's study who discovered that many students still failed to translate mathematical problems which denoted that the translation aspect among students was generally poor.

Regarding the interpretation aspect, most students were good at interpreting questions in the form of pictures into verbal form. Although some had not interpreted the answers completely or according to the concept. It was in line with the study of Amin et al [37] who stated that some students were not able to identify physical quantity variables and resulting in incomplete answers. Meanwhile, the result for the extrapolation aspect showed that most students were unable to describe the correct answers according to the concept. It was identified through the students' mistakes in determining the trajectory of the elevation angle. Most of the students could not predict symbols or graph slope patterns on questions [38]. These results contradict the research conducted by Susac et al [39] who found that most students have a good understanding of the extrapolation aspect. Students in their study successfully predicted the graphs from the mathematical form on kinematics material.

Due to the small sample study, the Mann-Whitney U-Test was used for the hypothesis test with a significance level of 0.05 criteria [40]. The results of hypothesis testing can be seen in Table 2.

Tabel 2. Hypothesis test

Class	U_{count}	u_{table} (α = 0,05)	Decision
Experiment	94.5	114	H_1 accepted
Control			

Based on Table 2, the value of u_{count} was 94.5, which was higher than the value of u_{table} of 114. The value of u_{count} was outside the reception area of H_0 which means that the H_0 was rejected and H_1 was accepted. It indicates that there was an increase in understanding of concepts in students who learned with direct instruction assisted by the online PhET simulation. This method has proven to be quite effective to foster students' interest and enthusiasm in learning [41]. In addition, this method enables students to have the opportunity to discuss and express their opinions related to questions in the worksheet based on their understanding using PhET simulations. Students can conclude the results of their discussions and can better understand and explain the concepts of physics that they have learned.

4. Conclusion

The results indicated that there was an effect of the direct instruction with PhET simulation in online learning on students' conceptual understanding of parabolic motion. It shows that PhET simulation is beneficial for students' learning outcome in both online and offline learning. Moreover, the use of PhET simulation is suitable to be integrated with direct instruction since this model allows the students to further discuss and restate the concepts they have learned on their own words through the peer discussion, and presenting and feedback session stage.

Acknowledgement
We highly acknowledge the support of the Program Kompetisi Kampus Merdeka (PK-KM) Grant of the Education, Culture, Research, and Technology of the Ministry of the Republic of Indonesia, with funding identity: 119/E1/KM.05.03/2021.

References
[1] Lestiyanawati R 2020 The Strategies and Problems Faced by Indonesian Teachers in Conducting e-learning during COVID-19 Outbreak *CLLiENT (Culture, Lit. Linguist. English Teaching)* **2** 71–82
[2] Whitacre I, Hensberry K, Schellinger J and Findley K 2019 Variations on play with interactive computer simulations: balancing competing priorities *Int. J. Math. Educ. Sci. Technol.* **50** 665–81
[3] Rahayu C D and Sartika S B 2020 Students Learning Motivation and Concepts Understanding of Science through the Use of PhET Interactive Simulations *SEJ (Science Educ. Journal)* **4** 63–76
[4] Sulisworo D, Ramadoan N, Ardianti S and Widodo 2019 Promoting PhET Simulation on Online Learning to Enhance Critical Thinking Skills *5Th Int. Conf. Educ. Res. Pract. 2019 Educ. Digit. Soc. Integr. Humanist. Sci. Values* 58–63
[5] Batuyong C T and Antonio V V 2018 Exploring the Effect of PhET ® Interactive Simulation-Based Activities on Students' Performance and Learning Experiences in Electromagnetism *Asia Pacific J. Multidiscip. Res.* **6** 121–31
[6] Buchori A, Setyosari P, Dasna I W, Ulfa S, Degeng I N S and Sa'dijah C 2017 Effectiveness of Direct Instruction Learning Strategy Assisted by Mobile Augmented Reality and Achievement Motivation on Students Cognitive Learning Results *Asian Soc. Sci.* **13** 137
[7] Habibbulloh M 2019 Effectiveness Of the Guided Discovery Model Based Virtual Lab Phet Toward Mastery Students' Concept On Topic Photoelectric Effect *Sci. Educ. Appl. ...*
[8] Faour M A, Ayoubi Z and The Z 2018 The Effect of Using Virtual Laboratory on Grade 10 Students' Conceptual Understanding and their Attitudes towards Physics *J. Educ. Sci. Environ. Heal.* **4** 54–68
[9] Taibu R, Mataka L and Shekoyan V 2021 Using PhET simulations to improve scientific skills and attitudes of community college students *Int. J. Educ. Math. Sci. Technol.* **9** 353–70
[10] Habibi H, Jumadi J and Mundilarto M 2020 Phet simulation as means to trigger the creative thinking skills of physics concepts *Int. J. Emerg. Technol. Learn.* **15** 166–72
[11] Astutik S and Prahani B K 2018 The Practicality and Effectiveness of Collaborative Creativity Learning (CCL) Model by Using PhET Simulation to Increase Students' Scientific Creativity *Int. J. Instr.* **11** 409–24
[12] Sarwoto T A, Jatmiko B and ... 2020 Development of online science teaching instrument based on scientific approach using PhET simulation to improve learning outcomes at elementary school ... *Int. J. ...*
[13] Perdana R, Riwayani R, Jumadi J and Rosana D 2019 Web-Based Simulation on Physics Learning to Enhance Digital Literacy Skill of High School Students *JIPF (Jurnal Ilmu Pendidik. Fis.* **4** 70
[14] Prima E C, Putri A R and Rustaman N 2018 Learning solar system using PhET simulation to improve students' understanding and motivation *J. Sci. Learn.* **1** 60
[15] Sari D P, Tjandrakirana T and Kuntjoro S 2018 Applying Science Learning Phet Simulation To Improve Process Skill And Knowledge Aspect Of Junior High School Student *JPPS (Jurnal Penelit. Pendidik. Sains)* **7** 1496
[16] Oktavia F, Muhibbuddin and Mursal 2020 Direct instruction model assisted with PhET and Whatsapp to students' cognitive and motivation *J. Phys. Conf. Ser.* **1460**
[17] Warju W, Ariyanto S R, Soeryanto S, Hidayatullah R S and Nurtanto M 2020 Practical Learning Innovation: Real Condition Video-Based Direct Instruction Model in Vocational

Education *J. Educ. Sci. Technol.* **6** 79–91
[18] Fayanto S, Misrawati M, Sulisworo D, Istiqomah H F N and Sukariasih L 2019 The Implementation of Multimedia on Physics Learning Based on Direct Instruction Model in The Topic of Light *Indones. J. Learn. Educ. Couns.* **1** 124–32
[19] Polonia B S E and Ravi A 2020 Effect of Direct Instruction Models Toward Students' Understanding of Physics Formula *Berk. Ilm. Pendidik. Fis.* **8** 133
[20] Bhuana G P and Apriliyanti D L 2021 Teachers' Encounter of Online Learning: Challenges and Support System *J. English Educ. Teach.* **5** 110–22
[21] Marín Díaz V, Reche Urbano E and Maldonado Berea G 2013 Advantages and disadvantages of online training *Ridu* **7** 2
[22] Hamid R, Sentryo I and Hasan S 2020 Online learning and its problems in the Covid-19 emergency period *J. Prima Edukasia* **8** 86–95
[23] Wuryanti S, Hadiana D and Purwati R 2020 Factors affecting the translation, interpretation, and extrapolation abilities in elementary school students' learning achievement *J. Penelit. dan Eval. Pendidik.* **24** 198–207
[24] Martín Ramos P, Ramos Silva M and Silva P 2017 Smartphones in the teaching of Physics Laws : projectile motion *RIED. Rev. Iberoam. Educ. a distancia*
[25] Serevina V and Raida R 2021 Improving the Quality of Education in the Covid-19 Era Through the Implementation of Online Learning Resources With Poe2We Model on Parabolic Motion *Int. J. Educ. Manag. Innov.* **2** 13
[26] Kantrowitz R and Neumann M M 2020 The English Galileo and His Vision of Projectile Motion under Air Resistance ed H Engler *Int. J. Math. Math. Sci.* **2020** 9695053
[27] Trudel L 2017 High School Students ' Conceptual Understanding Of Parabolic Motion *Hawaii University International Conferences*
[28] Mufit F 2018 The Study of Misconceptions on Motion's Concept and Remediate Using Real Experiment Video Analysis
[29] Mambu J Y 2020 Based Physics Game on Android **6** 239–50
[30] Pineda C I S 2020 Effectiveness of Validated Teaching-Learning Package in Projectile Motion for Grade 9 Science *J. Sci. Sci. Educ.* **1** 26
[31] Rohadi N, Setiawan I and Swistoro E 2019 Descriptive Qualitative Analysis Based on Physics Mental Model of Students' Cognitif Understanding on the Concepts of Kinematics Particle and Parabolic Motion *J. Pembelajaran Fis.* **8** 121–8
[32] Minium E W, King B M and Bear G 1993 *Statistical Reasoning in Psychology and Education*
[33] Salame I I and Makki J 2021 Examining the Use of PhET Simulations on Students' Attitudes and Learning in General Chemistry II *Interdiscip. J. Environ. Sci. Educ.* **17** e2247
[34] Özcan H, Çetin G and Koştur H İ 2020 The Effect of PhET Simulation-based Instruction on 6th Grade Students' Achievement Regarding the Concept of Greenhouse Gas *Sci. Educ. Int.* **31** 348–55
[35] Inayah, Nailil,Masruroh M 2021 PhET Simulation Effectiveness as Laboratory Practices Learning Media to Improve Students ' Concept Understanding *Prism. Sains J. Pengkaj. Ilmu dan Pembelajaran Mat. dan IPA IKIP Mataram* **9** 152–62
[36] Jelatu S and Kurnila V S 2019 Analysis of Translation Understanding from Verbal to Visual on Trigonometry Concept *Form. J. Ilm. Pendidik. MIPA* **9** 191–202
[37] Amin B D, Sahib E P, Harianto Y I, Patandean A J, Herman and Sujiono E H 2020 The interpreting ability on science kinematics graphs of senior high school students in South Sulawesi, Indonesia *J. Pendidik. IPA Indones.* **9** 179–86
[38] Veloo A, Krishnasamy H N and Wan Abdullah W S 2015 Types of student errors in mathematical symbols, graphs and problem-solving *Asian Soc. Sci.* **11** 324–34
[39] Susac A, Bubic A, Martinjak P, Planinic M and Palmovic M 2017 Graphical representations of data improve student understanding of measurement and uncertainty: An eye-tracking study *Phys. Rev. Phys. Educ. Res.* **13**

[40] Corder, Gregory W and Foreman D I 2014 *Nonparametric statistics: A step-by-step approach* (ohn Wiley \& Sons)
[41] Haryadi R and Umam A K 2021 Pemanfaatan Simulasi Phet Dalam Mendukung Pembelajaran Fisika Materi Rangkaian Listrik Searah Pada Masa Pandemi *Edufisika J. Pendidik. Fis.* **6** 18–21

Journal of Physics: Conference Series **2126** (2021) 012014 doi:10.1088/1742-6596/2126/1/012014

Students' scientific representation enhancement through "ECRA" techniques in formative examination design

Linda Nurul Jannah Mohd Rizal Khoo[1*], Nor Farahwahidah Abdul Rahman[1], Salmah Othman[2], Noraidah Sobri[2]

[1]School of Education, Faculty of Social Sciences and Humanities, Universiti Technologi Malaysia, 81310 Skudai, Johor.
[2]Melaka Matriculation College, Londang, 78300, Melaka.

[*]Corresponding author: jannah.hlp@gmail.com

Abstract. The recent studies in science learning has a specific interest with 4C's which are communication, collaboration, critical thinking and creativity. This study explored the students' 4C's as they engage with physics concepts relating to alternating currents. This engagement entails students' interpretation for descriptive questions. Following action research cycle, the study begin with planning the intervention using "ECRA" technique and conducted the activities with 38 Matriculation College students. The analysis attempt to make a connection between students' ability to construct representational aspects of scientific description with formative achievement. This study systematically describes students' understanding on the representations signify in the descriptive questions. Data collected from this study were Work Report, review of answers tutor questions, and interviews during observation stage. The finding suggests that, the ECRA technique is enabling students to obtain higher percentage in formative assessment by allowing them to developed richer conceptual understanding. Hence this study suggest that the representational issues in physics questions constructed in essay can be moderated with ECRA technique.

1. Introduction

Physics is a lesson about facts, concepts, theories, principles and natural laws [1]. When learning Physics, students' should not only memorize what they had learnt in the class, instead learning must revolves around doing experiment and Physics concepts comprehension. Nowadays, emphasis on 21th century invite more attention to four main competency skills required as learning outcomes. In the end of the lesson, students must acquire critical thinking skills, creative thinking skills, communication skills and collaborative skills [2]. These skills also known as 4C competency.

Critical thinking is a fundamental skill in problem solving. Critical thinking should be not only an educational option but an integral part of education practices[3]. Critical thinking skills is needed to foster students' thinking skills since it requires examining assumptions, discerning hidden values, evaluating evidence, and assessing conclusions[4]. Students' ability to interpret the descriptive questions are important to ensure that students are able to translate the different representatives used in the exam questions. Therefore problem solving although its integral to knowledge development, students must be guided to ensure the process meet the answers scheme demand. Guiding students for this standard require an introduction to a proper answering technique to accompany the problem-solving process. One of the thinking skills that is expected to be emerge in physics learning is critical thinking skills.

Critical thinking can stimulate students to solve the problems that are related to the learning materials[5]. The better development of these abilities will make students' able to overcome the problems with satisfying results. Critical thinking can stimulate students to solve the problems that are related to the learning materials [5].

Similarly, creative thinking skill has its significant role when solving difficult physics concept. Therefore, learning requires a strategy or to ensure the creativity is visible to learners. Creativity is known for allowing higher reasoning abilities, stronger thinking abilities, increased level of trust, motivation and learning involvement, critical thinking and problem-solving abilities[6]. These outcomes can be obtained since creativity encourage students to propose various ideas, various solution approaches, offering new alternatives, and being able to explain easily [7]. This study aimed at improving student's attainment in examination by introducing specific answering technique that benefit students' creativity and critical thinking known as "ECRA" technique.

Muhammad Noor Kholid [8] finds that a non-routine question can cause the students have reflective thinking about the question given and lead to their confusion. The unfamiliar question for students will encourage them to apply their knowledge and experience to solve the problem. A non-routine question to explore the students' critical thinking and creativity in solving descriptive question. Accordingly, this study proposes a method for answering descriptive question in the Physics assessment.

2. Context of study

This study is an extension work from analyzing students' performance in final exam. One of the most concerning part is related to student's performance involving descriptive questions. From the Candidate Work Report (LKC) students did not get full marks for descriptive questions. In fact, students are able to get full marks in this descriptive question if they know the technique appropriate in answering the questions. A series of interviews were conducted to identify the possible reasons behind this outcomes. Students also complained of always not successful to get full marks for descriptive questions. The results of interviews with students showed that:

1. They are not interested to try and they try to avoid from answering descriptive questions.
2. They do not have the confidence and insist for right answering technique for descriptive question
3. Their physics concepts is weak.
4. They do not know the intent of this question and what needs to be stated in the answer.
5. The narrative of questions is not easy to be understood thus limit their chances to get full marks.

Apart from the interview sessions, a review of the tutoring and students' test answers showed students are

1. Not proficient in answering descriptive questions.
2. Students also tend to answer descriptive questions as a final choice of their trial to get the best results.
3. The test results also found that it was difficult for students to get marks full of descriptive questions. Usually full marks for such questions is 3 or 4 marks, but most students get only 1 and 2 marks.

3. Methods

This type of research was conducted using Descriptive Research Design (DRD). The results of this study were to obtain from the implementation of ECRA techniques in the pre-test and post-test. This research was conducted in February 2019 in Malacca Matriculation College, specifically in Module I students. The subjects of this study were 38 students of SDS Matriculation Programme Session 2019/2020. Data collection methods by using SPSS and questionnaire given to the students. The research instrument has been validated by two experts and has been declared valid and reliable.

Following the prior analysis, this study attempts to introduce students with a technique for answering descriptive questions to overcome the problems faced by students. The technique employed as an intervention is the "ECRA" technique. ECRA is an acronym represented by Equation, Constant variable,

Journal of Physics: Conference Series **2126** (2021) 012014 doi:10.1088/1742-6596/2126/1/012014

Relationship and Answer. The abbreviation describes the steps needed to be taken when answering descriptive question. Therefore, a quasi-experimental study without controlling group is used as the methodology for this study. The students' improvement by applying this skill in answering descriptive questions were measured through the percentage of achievement students' pass, increasing grade A and full marks in the descriptive questions.

The study conducted is focused on improving students' problem-solving skills for Matriculation Physics subjects in alternating current chapter. There were 4 activities were carried out in this research. Here are the steps of the activities that have been implemented.

3.1. Activity 1: Pretest

For Activity 1, students are required to answer a pretest question which contains 1 descriptive question. The question and the sample of student' answer are presented in Figure 1 and 2, respectively. Pre-tests were conducted for 10 minutes.

LAMPIRAN A

NAME: **PRACTICUM:**

UJIAN PRA

An ac generator with a frequency equal to the resonance frequency of the circuit is connected in series with an inductor of inductance 75 mH, a capacitor of capacitance 1 mF and a resistor of resistance 100 Ω. If both inductance and capacitance are doubled, does the frequency increase, decrease or remain constant? Explain.

Figure 1. Pretest descriptive question

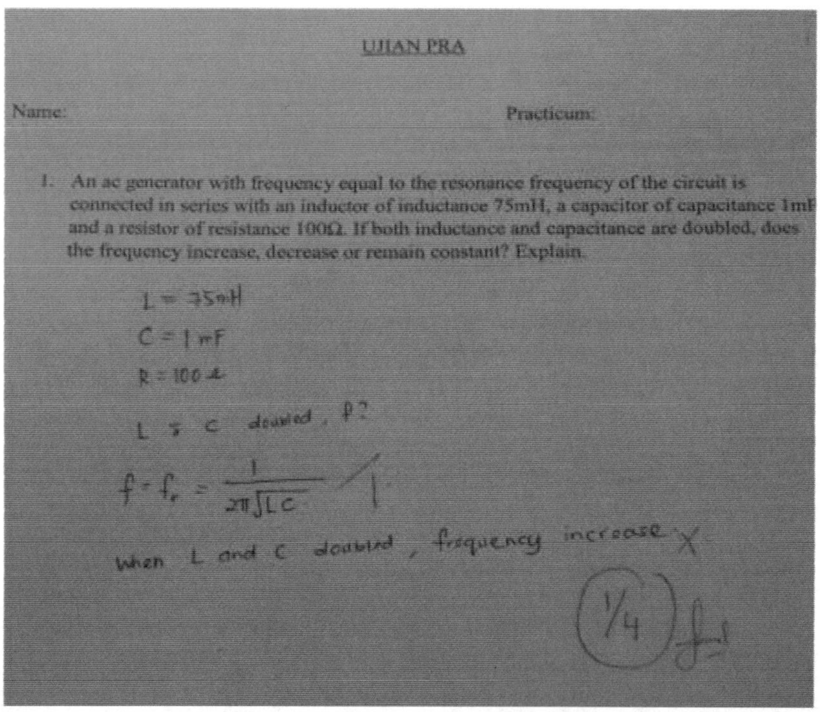
UJIAN PRA

Name: Practicum:

1. An ac generator with frequency equal to the resonance frequency of the circuit is connected in series with an inductor of inductance 75mH, a capacitor of capacitance 1mF and a resistor of resistance 100Ω. If both inductance and capacitance are doubled, does the frequency increase, decrease or remain constant? Explain.

$L = 75 mH$
$C = 1 mF$
$R = 100 Ω$
L & C doubled, f?
$f = f_r = \frac{1}{2\pi\sqrt{LC}}$
When L and C doubled, frequency increase ✗
1/4

Figure 2. Sample of student answers

Figure 2 shows an example of students' answers in a given pre -test. Next, the lecturer will check the students' answers and the marks are recorded.

3.2. Activity 2: Implementation of "ECRA" technique
After carrying out Activity 1, Activity 2 was implemented to all students. In this activity the "ECRA" technique is introduced. Students are given an explanation about steps need to be taken into account when answering the descriptive questions using the guidance words E: Equation, C: constant variable, R: relationship and A: Answer.

3.3. Activity 3: Posttest
For activity 3, students were given a posttest which contains 2 descriptive questions that need to be answered within 7 minutes. Question given is similar to the question in the prettest. Next, the lecturer will check post-test students' answers of scores were recorded for both groups.

3.4. Activity 4: Questionnaire
The lecturer distributed the questionnaire to all students. The goal is to obtain student feedback on the use of the "ECRA" technique. There are 7 items that should be answered in this questionnaire (see Table 1). The questionnaire survey was designed to meet the goal of this study based on items concerning the self-efficacy, critical thinking intrinsic value, creative thinking and collaborative skills.

Table 1. Items in Questionnaire survey

Name	Description	Answer form
Self-efficacy: within the scope of the course	The use of 'ECRA' techniques is very easy to understand.	5-point scale assessment 1:Strongly disagree, 2: Disagree, 3: Undecided, 4: Agree 5: Strongly agree
Self-efficacy: save time	The use of the "ECRA" techniques save my time in answering formative questions.	
Critical thinking: understand	The use of the "ECRA" techniques helped me answering the formative question correctly.	
Intrinsic value: interesting	I am more enthusiastic about answering descriptive questions using the "ECRA" techniques.	
Creative thinking: answering method	The use of "ECRA" techniques helped me from making mistakes while answering test questions.	
Intrinsic value: improvement	The use of ECRA techniques improved my performance in answering test questions.	
Collaborative skills : introduced to other students.	"ECRA" techniques should be introduced to all students.	

4. Results and Discussion
Figure 6 shows the comparison of scores between pre-test and post-test. There are as many 38 students participated in this study. For both pre-test and post-test, 5 marks and below consider as fail. In the Pre-Test a total of 8 student failed in this test while in the post-test no student failed. While the number of students who passed the pre –test was 28 people and 38 students passed in post-test. The number of students who got grade A, a total of 8 people got grade A's in pre-test and it has increased to 31 people in post-test. For the number of students who got full marks, only 6 people got full marks in the pretest and has increased to 25 people in the post-test. In addition, a total of 32 students managed to improve the score in the posttest. The detailed comparison of scores between pretest and posttest is presented in Figure 3.

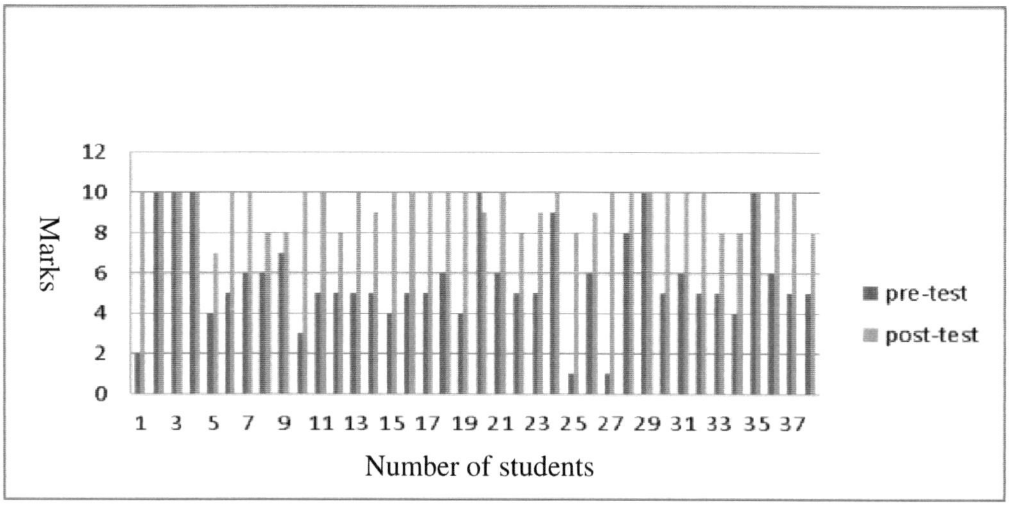

Figure 3. Comparison of scores between pre-test and post-test

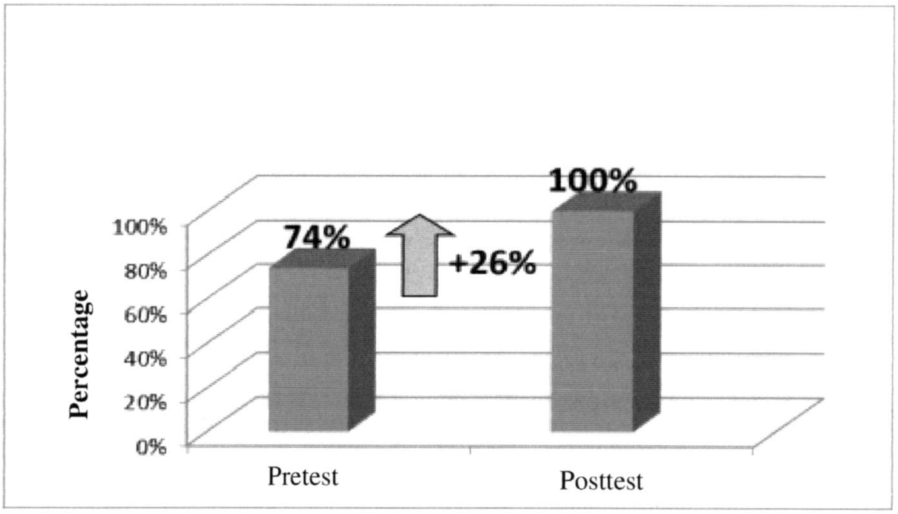

Figure 4. Comparison of percentage of students get pass between pre-test and post-test

Figure 4 shows a comparison of the percentage of students get pass for both the pre-test and post-test. For the pre-test, 74% of students passed in this test. While for the post-test has increased to 100% of graduating students. Therefore this study found that there have increase in the percentage of students passed the test by 26%.

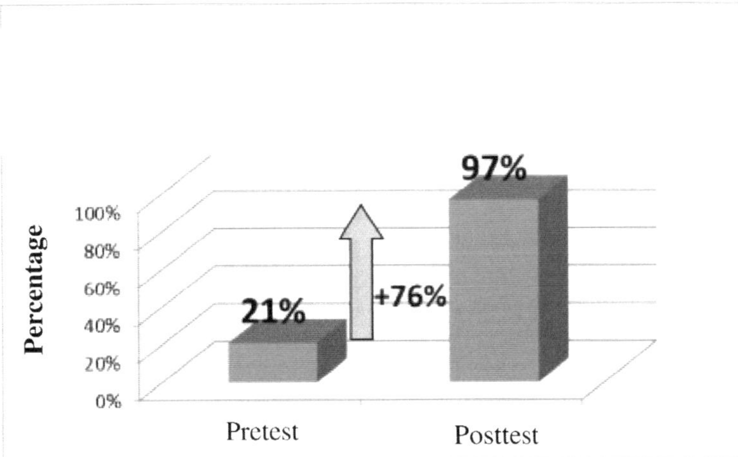

Figure 5. Comparison of the percentage of students getting A's between pre-test and post-test

Figure 5 shows a comparison of the percentage of students getting grade A's for the both tests. For the pre-test, 21% of the students got an A grade in this test. Meanwhile, for the post-test, it has increased to 97 % of students getting an A grade. This study found that there was an increase in the percentage of students getting an A grade by 76%.

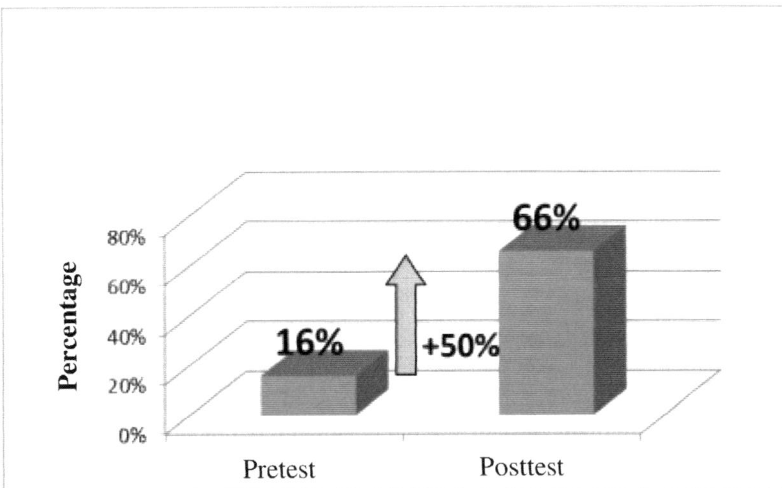

Figure 6. Comparison of the percentage of students getting full marks.

Figure 6 shows the comparison of the percentage of students getting full marks for both pre-test and post-test. For the pre-test, 16% of the students got full marks in this test. Meanwhile, for the post-test, it has increased to 66 % of students got full mark. This study found that there was an increase in the percentage of students getting marks full by 50%. Number of students who passed, got full marks and got A's grade for pre-test and post-test as well as the number of students who have an increase in marks in the posttest.

Following are the findings from the questionnaire that has been implemented to a total of 38 respondents. All questionnaire findings were analyzed and each item have be calculated the average scores are shown in Table 2.

Table 2. Mean score values for questionnaire items

No	Item	Average Score
1.	The use of 'ECRA' techniques is very easy to understand.	4.59
2.	The use of the "ECRA" techniques save my time in answering formative questions.	4.47
3.	The use of the "ECRA" techniques helped me answering the formative question correctly.	4.59
4.	I am more enthusiastic about answering descriptive questions using the "ECRA" techniques.	4.56
5.	The use of "ECRA" techniques helped me from making mistakes while answering test questions.	4.68
6.	The use of ECRA techniques improved my performance in answering test questions.	4.68
7.	"ECRA" techniques should be introduced to all students.	4.68
	AVERAGE	4.61

Based on Table 2, it was found that all the questionnaire items obtained a mean score above 4.00 and an average mean score is 4.61 which describes the use of the technique "ECRA" is at a level of very satisfying. The highest mean score value was 4.68 which was recorded on the 3 items. Based on the highest score values of these three items prove that the "ECRA" technique helps improve students 'skills of answering descriptive question.

The instrument of the study comprised: 1) a test with a questions related to alternating current, 2) a note of introduction of ECRA technique for understanding the students' descriptive question problem solving, and 3) a questionnaire for confirming the students' ability to solve the descriptive type of question. Referring the results of the data analysis above, it can be stated that the students of Malacca Matriculation, Malaysia tend to use the ECRA technique in solving descriptive question for alternating current. Most of the students play their role in creative thinking and they understand the purposes to answer the questions correctly. It is agreed by the Handayani et.al that the students understand the purpose of the problem solving[9]. In the monitoring component, researchers found that they have improvement in term of getting grade A's, pass and full marks in the assessment.

Based on the results of the test and questionnaire by the students, introducing the ECRA techniques improves the students' problem solving skills. This problem solving skills helps teachers to increases students motivation, students learn how to organize themselves to learn physics, and increase the ability to think in creative way to answers the questions given in the assessment.

5. Conclusion

Following are the marks obtained for the pre-test and post-test. The results show that all students that involved in the study have passed 100% after being introduced to the "ECRA" technique. It is found that the percentage of students who get grade A's has also increased by 76% from 21% in the pre-test and to 97% in the post-test. The increase in the percentage of grade A's has exceeded the objective target of the study which is to increase the percentage students who get an A grade of 50%. Percentage of students who get full marks has increased by 50% from 26% to 66% after "ECRA" was introduced and this achievement meets the target of the study which is to increase the percentage of students who get full as much as 50%.

Findings from the student responses in the questionnaire have also shown that this techniques helps to improve students' skills in answering descriptive questions. This study was conducted after students learned the topic of "Alternating Current". There are many descriptive questions in this topic. After the introduction of the "ECRA" technique to this group of students, researcher found that this technique can improved students' skills in answering descriptive questions.

Acknowledgement
This project is funded under Fundamental Research Grant Scheme (FRGS) FRGS/1/2018/SSI09/UTM/02/5 by The Ministry of Education Malaysia with vote number R.J130000.7853.5f067.

References
[1] Hardiyanto H, Susilawati S and Harjono A 2017 Pengaruh Model Pembelajaran Berbasis Masalah dan Ekspositori dengan Keterampilan Proses Sains Terhadap Hasil Belajar Fisika Siswa Kelas VIII MTsN 1 Mataram Tahun Ajaran 2014/2015 *J. Pendidik. Fis. dan Teknol.* **1**
[2] Supena I, Darmuki A and Hariyadi A 2021 The influence of 4C (constructive, critical, creativity, collaborative) learning model on students' learning outcomes *Int. J. Instr.* **14**
[3] Zeng H, Zhou S N, Hong G R, Li Q Y and Xu S Q 2020 Evaluation of interactive game-based learning in physics domain *J. Balt. Sci. Educ.* **19**
[4] Anon 2018 Gamification Approach in Education to Increase Learning Engagement *Int. J. Humanit. Arts Soc. Sci.* **4**
[5] Saputra M D, Joyoatmojo S, Wardani D K and Sangka K B 2019 Developing critical-thinking skills through the collaboration of Jigsaw model with problem-based learning model *Int. J. Instr.* **12**
[6] Richardson C and Mishra P 2018 Learning environments that support student creativity: Developing the SCALE *Think. Ski. Creat.* **27**
[7] Calderón M, Slavin R E, Sánchez M, Calderón M, Slaviti R, Sánchez M, et al 2015 Learner-based teaching systems for English for specific purposes. *Teach. Teach. Educ.* **63**
[8] Kholid M N, Sa'dijah C, Hidayanto E and Permadi H 2020 How are students' reflective thinking for problem solving? *J. Educ. Gift. Young Sci.* **8**
[9] Handayani A D, Herman T, Fatimah S, Setyowidodo I and Katminingsih Y 2018 Inquiry based learning: A student centered learning to develop mathematical habits of mind *Journal of Physics: Conference Series* vol 1013

The analysis of student kinesthetic learning activity on the materials of Compton and photoelectric effects

S Saehana[1], I K Werdhiana[1*], N S Safitri[1], O Saputra[1], and N Safira[1]

[1]Sub-Department of Physics Education, Faculty of Teacher Training and Education, Universitas Tadulako, Jl. Soekarno Hatta KM. 9 Kampus Bumi Tadulako Tondo Palu-Central Sulawesi, Indonesia.

*Corresponding author: ikomangwerdhiana@yahoo.co.id

Abstract. This study aims to analyze the kinesthetic learning activities of students on Compton and photoelectric effects. This further involved a qualitative descriptive study, where students of class XII Mathematical and Natural Science 2, State of Madrasah Aliyah in Palu Region (Central of Sulawesi, Indonesia), were the sample population examined, in the odd semester of 2018/2019 academic session. Data was obtained through observation of learning activities, concept understanding tests, interviews and learning style identification questionnaires. The selection of respondents was based on the level of understanding of students' concept categories. The results of the study showed that students perform kinesthetic learning activities very well, as seen from their endeavors in following the lessons and the results of theoretical understanding assessments, which were observed to be in the high category. Furthermore, the interview evaluated that the students are happy with the kinesthetic learning method carried out as it promoted their understanding and recollection of the content of the material and questionnaires indicate that students have visual, auditory and kinesthetic learning styles. From the performance of Compton and the photoelectric effect, it was shown that students playing as electrodes and electrons (kinesthetic learning style) possess high understanding of the concept in the kinesthetic learning model.

1. Introduction

The learners with different learning styles have different sets of learning sequences [1]. Every student has their personalized pattern of absorbing and processing information received [2], as one with a visual learning style finds it easier to understand from what is seen. Furthermore, a person with auditory cleverness finds it easier to learn from what is heard, however, kinesthetic learning patterns are observed in someone who can easily mentally grasp information from motion and touch [3].

Based on the students' preferred technique, tutors can further adjust their teaching methods to suit their needs. The optimal learning requires that students receive instruction tailored to their putative learning style [4]. This process however involves the knowledge of the style of students as they must master various instruction formats and further be capable of being experts in the materials provided, motivators and sources of information, instructors and evaluators [5].

Motivation is needed for students to have a better and effective understanding of the material presented, by making learning activities interesting, through the involvement of a kinesthetic approach [6]. Furthermore, this education pattern increases student motivation to take lessons, decreases boredom and they are better able to understand the material given [7].

Kinesthetic Learning Activities (KLA) is a process that involves more physical actions than passive listening, with the main characteristic, being the active use of physically participating in material presentations in class [8]. KLA supports increased understanding of student concepts, encourages them to learn in new ways, increases their involvement and participation [9], and it is also fun and memorable and they inject a great deal of excitement and enthusiasm into the classroom [10].

In studying physics in class, there are several phenomena that can be seen with the eye. However, there are also many phenomena that cannot be seen with the eye so that students may experience difficulties in learning which lead to a lack of understanding of concepts and can even cause students to experience misconceptions [11,12]. Based on this, of course a teacher must implement learning that can facilitate students in learning so that students' conceptual understanding is good. One way for teachers to overcome these problems is by implementing KLA in learning whose concepts are difficult for students to understand.

Several studies have implemented KLA. The drawback of this research is that it applies KLA to concepts that can be observed by students [13–15]. There are still a few researchers who conduct research on the implementation of kinesthetic learning on material whose concepts cannot be seen by students directly. One of the concepts that cannot be observed by the human eye is compton and photoelectric effects.

Compton and photoelectric effects are interesting phenomena in modern physics which became the cornerstone of its development and it is also a part of the dualism of particle waves. Furthermore, both phenomena take place in a microscopic order hence they cannot be directly observed. The process of learning physics requires that teachers generally use images or animations, while presenting the materials, however, not all students can understand the explanation, because each student possesses a preferential learning style [16].

In this study, KLA was applied to the Compton effect and the photoelectric effect materials, aimed at improving understanding, where students demonstrated the behavior of microscopic objects, to facilitate learning of the study material.

2. Methods

This is a descriptive qualitative research, which describes students' KLA. The sample in this research is grade 11 students (children with an age range of 18-19 years) majoring in science at a public school in Central Sulawesi, Indonesia, during the odd semester of 2018/2019.

Respondents consisted of 17 students who had never studied Compton and photoelectric effect, where each was presented with different learning styles. Furthermore, the selection of respondents was based on level of understanding the concept and the differences in learning styles, based on the level of understanding of students' concepts will be grouped into three categories, namely high category, medium category and low category. Two respondents will be taken from each category so that a total of 6 respondents will be interviewed and given a learning style questionnaire until the learning style of each respondent is obtained.

The research was conducted by providing material on the working principle of both topics in physics, accompanied by a demonstration (kinesthetic learning activities) and a test was administered. The selected respondents were interviewed and given questionnaires.

The test consisted of 5 items on the principle of the Compton and photoelectric effects and the interviews related to the materials as well as information on the respondent's learning pattern. Furthermore, the questionnaire consisted of 18 items consisting of several statements about student learning styles, with scores determined by the Guttman scale, where each "YES" answer was graded 1 and the "NO" answer was 0 [17].

Determination of student learning styles were as follows:

1) The highest score in a group of learning style statements shows the dominance of the learning style.
2) The two peak values and equivalent records from both groups of learning style statements indicates the dominance of the two learning styles

3. Results and Discussion

The photoelectric effect and the Compton Effect are materials that are difficult for students to understand because they are abstract phenomena that cannot be observed directly, hence students need to use their imagination and utilize learning media.

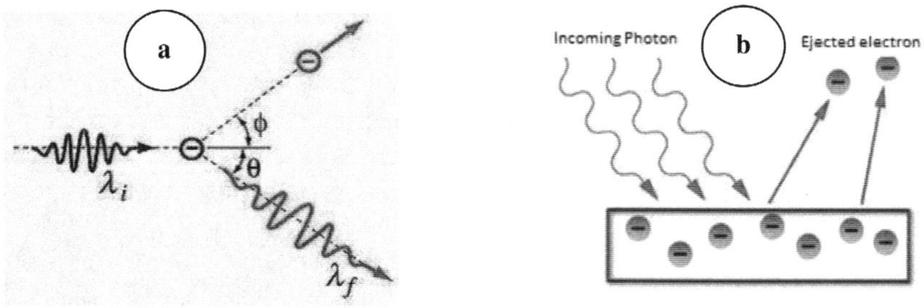

Figure 1. Concept: (a) Compton Effect and (b) photoelectric effect.

Figure 1 illustrates two phenomena in physics, where the Compton Effect event shown in Figure 1 (a), shows that photons that move with a certain wavelength hit idle electrons, which bounce hence the photons scatter with a larger wavelength and the difference between the wavelength of the initial and scattered photon is shown in Equation 1.

$$\lambda' - \lambda = \frac{h}{m_e c}(1 - cos\ cos\ \theta) \quad (1)$$

Where:
λ' is the initial wavelenght,
λ isthe scattered wavelenght,
h is the Planck's constant,
m_e is the idle mass of electrons,
c is the speed of light, and
θ is the photon scattering angle relative to the flat line.

Conversely, the photoelectric effect event, as shown in Figure 1 (b), illustrates a relationship between several physical parameters as shown in Figure 2. where a higher intensity of the photon that hits the metal plate, generates a greater electric current as shown by Figure 2 (a). Furthermore, the electrons on the metal plate are further excited if the photon energy involved is greater than the metals threshold energy as shown in Figure 2 (b).

Figure 2. Shows (a) the relationship between physical parameters in the photoelectric effect event and the effect of photon intensity on the electric current produced as well as (b) the correlation between the frequency and kinetic energy of electrons.

At the time of the implementation of learning in accordance with the student activity observation sheet there are 5 stages. The first stage is paying attention and listening to explanations about the learning objectives, noting the learning objectives presented, understanding the problems presented, and students' skills in answering questions. The second stage, in groups according to the results of group division, paying attention to explanations, noting the points of learning that are delivered, asking for explanations that are not yet clear. The third stage, draw the lottery to find out the role played, students actively discuss in their respective groups discussing the role they have got, students prepare to play a direct role as part of the Compton effect and the photoelectric effect, students use the facilities provided. The fourth stage, another group evaluates the group that appears, students focus their attention on the role played and do not do things that are not related to procedures, students make movements according to their role, the next designated groups are ready to appear. The fifth stage, reading out the assessment of the performance of each group, giving conclusions when asked, writing the conclusions that have been concluded, doing a test of understanding the concept of the Compton effect and the photoelectric effect.

Using KLA, students performed two demonstrations, where some roles in exhibiting the Compton Effect, involved photons arrival, idle electrons, scattered photons, and bounced electrons. Conversely, the photoelectric effect exhibition required that students act as photons, metal plates, electrons in metal, and electrodes (both anodes and cathodes).

The results show that students demonstrated both physical phenomena very well, with data obtained through the observation of their activeness, during the course of the lesson, at a participatory percentage of 95%. Furthermore, their KLA was observed based on the roles played, utilizing indicators, with a purpose to find out the suitability of the students' movements, while acting out their characters.

The kinesthetic indicators assessed, include locomotor skills (walking and marching), nonlocomotor skills (turning the body, standing up and reaching out), eyes and foot coordination (ability to manipulate objects), as well as the ability to control and regulate the body (take stars and change direction). Furthermore, students will be given a concept understanding test after kinesthetic learning is applied. Based on data analysis, the average score obtained by students is 69.7 or categorized in the medium category. Category grouping further used the standard deviation [18].

The categorized students, from the results of the calculation of standard deviations as seen in Table 1, were used as the basis for grouping students, where two respondents were selected from each category to be interviewed and further administered a learning style questionnaire to obtain the data in Table 2.

Table 1. Score of Respondent Category Determination

	High	Moderate	Low
Score	$\geq 78{,}35$	$78{,}35 < x < 61{,}05$	$\leq 61{,}05$

In the Compton effect demo, students who acted as photons, energy and colliding electrons were more active in kinesthetic activity because they carried out more physical movements and thus fulfilled all indicators, as shown in Figure 3 (a).

Figure 3. Student kinesthetic activity on Compton effect: (a) The role of students as electrons,

Journal of Physics: Conference Series **2126** (2021) 012015 doi:10.1088/1742-6596/2126/1/012015

energy and photons; (b) The role of students as silent electrons and x-rays.

However, the opposite occurred when students acted as x-rays and idle electrons, where they tend to be more passive in kinesthetic activities, as shown in Figure 3 (b), because little movements were required. In line with the results obtained at the demonstration of the Compton Effect, it is known that students, who acted as photons and electrons according to the photoelectric material, were more active in kinesthetic activity because they made more physical actions and thus fulfilled all kinesthetic indicators (Figure 4 (a)).

Figure 4. Student kinesthetic activity in the demonstration of the photoelectric effect: (a) The role of students as electrons and photons; (b) The role of students as light sources, anodes and ketodes.

In the photoelectric effect, students who acted as light sources, anodes and cathodes, were more passive in kinesthetic activity, as it is known through the indicators that only made a few movements (Figure 4 (b)).

The results further indicated that the level of understanding of the respondents on the working principle of Compton and photoelectric effects were influenced by the role played as it was observed that those who obtained active roles, e.g. colliding electrons, energy, and photons tend to possess a better understanding and were able to explain the working principles of Compton effects and photoelectric effects. In contrast, respondents who obtained the passive roles, e.g. X-rays, light sources, cathodes, anodes and idle electrons did not have a good understanding of the concept of both phenomena and they also possessed lower explanatory abilities.

The results of the identification of learning styles and categories or levels of students' conceptual understanding after studying by the demonstration of compton and photoelectric effects are shown in Table 2. These respondent were representatives for each category in learning styles. It is known that respondents with kinesthetic learning styles (R-07), obtained high learning outcomes because in the demonstration of both physics phenomena, respondents (R-07) obtained a passive role (as X-rays and light sources) sequentially, where they only focused on their character.

Table 2. Respondent, categories, and learning styles test results

Respondent	Score	Category	Learning style
R-07	95	High	Kinesthetic
R-06	80	High	Auditorial Kinesthetic
R-09	70	Moderate	Auditorial
R-04	65	Moderate	Visual
R-15	55	Low	Auditorial
R-05	60	Low	Visual

It is also known that those who have a kinesthetic learning style (R-06 and R-07) used the graphical potential in their learning activities, hence, when playing their roles, they observed objects, figures and symbols around them, thus, the information was easier to obtain. Table 2 also shows that students with Kinesthetic Learning Styles and its combined resulting better understanding of the concept after implementing KLA than others student. This is in line with the research that has been done which states that students' understanding of concepts will increase if students are interested and demonstrate the material to be studied [19]. This research is also in line with previous research which states that if you want to improve students' understanding skills, it is not enough to learn in a passive style, but it must be kinesthetic (involving the movement of all limbs) [20].

Through role playing in KLA, it is known that students were happy to learn physics. KLA's can engage other important learning styles, such as Felder and Silverman's active, sensing, intuitive, visual, or global learners [21]. Furthermore, fun learning attracts students' attention because the various methods applied in the education process are interesting [22].

4. Conclusion

Students participated in excellent kinesthetic learning activities, which led them to be more active and happy, determined by observing their ability to follow the lessons, test results and interviews. The application of KLA, helped the students to understand and remember the material (shown in the learning style questionnaire). In kinesthetic learning, respondents with visual knowledge styles were in the high category because they have an active role hence understanding the working principle of the Compton Effect and the photoelectric effect was enhanced.

Acknowledgements

This work was supported by Penelitian Terapan Unggulan Perguruan Tinggi Research Grant from the Research, Technology and Higher Education of the Ministry of the Republic of Indonesia in the fiscal years of 2019. Also, we would highly acknowledge the support of the Program Kompetisi Kampus Merdeka (PK-KM) Grant of the Education, Culture, Research, and Technology of the Ministry of the Republic of Indonesia, with funding identity: 119/E1/KM.05.03/2021.

References

[1] Klašnja-Milićević A, Vesin B, Ivanović M and Budimac Z 2011 E-Learning personalization based on hybrid recommendation strategy and learning style identification *Comput. Educ.* **56** 885–99

[2] Prithishkumar I J and Michael S A 2014 Understanding your student: Using the VARK model *J. Postgrad. Med.* **60**

[3] Lista L, Atmowardoyo H and Salija K 2016 The Effects of Visual Auditory Kinesthetic Learning Style as Technique in Improving Students' Writing Ability *ELT Worldw. J. English Lang. Teach.* **2** 62–76

[4] Pashler H, McDaniel M, Rohrer D and Bjork R 2008 Learning styles concepts and evidence *Psychol. Sci. Public Interes. Suppl.* **9** 105–19

[5] Zapalska A and Brozik D 2006 Learning styles and online education *Campus-Wide Inf. Syst.* **23** 325–35

[6] Grant C and Osanloo A 2014 Understanding, Selecting, and Integrating a Theoretical Framework in Dissertation Research: Creating the Blueprint for Your "House" *Adm. Issues J. Educ. Pract. Res.* **4** 12–26

[7] Suryadin, Merta I W and Kusmiyati 2017 Pengaruh Model Pembelajaran Visual Auditorial Kinestik (VAK) terhadap Motivasi dan Hasil Belajar IPA Biologi Siswa Kelas VIII SMP Negeri 3 Gunungsari Tahun Ajaran 2015/2016 *J. Pijar MIPA* **12** 19–24

[8] Gashgari A and Young G 2017 A Study of Kinesthetic Learning Activities in Teaching Computer Algorithms in Saudi Arabia *Int'l Conf. Frontiers in Education: CS and CE*

[9] Mobley K and Fisher S 2014 Ditching the Desks: Kinesthetic Learning in College Classrooms

Soc. Stud. **105** 301–9

[10] Richards A 2019 Teaching Mechanics Using Kinesthetic Learning Activities *Phys. Teach.* **57** 35–38

[11] Saputra O, Setiawan A, Rusdiana D and Muslim 2020 Analysis of students' misconception using four tier diagnostic test on fluid topics *Int. J. Adv. Sci. Technol.* **29** 1256 – 1266

[12] Saputra O, Setiawan A, Rusdiana D, Muslim and Izzuddin M A 2021 The development of 5E-POW (predict-observe-write) learning model assisted by virtual simulation to reduce the quantity of high school students' misconceptions on fluid topics *Journal of Physics: Conference Series* vol 1806

[13] Whitworth B A, Chiu J L and Bell R L 2014 Kinesthetic Investigations in the Physics Classroom *Phys. Teach.* **52** 91–93

[14] Ganesh P 2020 Connecting Torque and Friction Using Tactile Learning Activities *Phys. Teach.* **58** 659–66

[15] Richards A 2020 Teaching Electricity and Magnetism Using Kinesthetic Learning Activities *Phys. Teach.* **58** 572–6

[16] Gilakjani A P 2012 A Match or Mismatch Between Learning Styles of the Learners and Teaching Styles of the Teachers *Int. J. Mod. Educ. Comput. Sci.* **4** 51–60

[17] Sugiyono 2015 *Quantitative Research Methods, Qualitative, and R & D.* (Bandung: Alfabeta)

[18] Sudjana 2005 *Statistical Methods* (Bandung: Tarsito)

[19] Ozkan G and Selcuk G S 2016 Facilitating conceptual change in students' understanding of concepts related to pressure *Eur. J. Phys.* **37** 1–20

[20] Leinonen R, Asikainen M A and Hirvonen P E 2013 Overcoming students' misconceptions concerning thermal physics with the aid of hints and peer interaction during a lecture course *Phys. Rev. Spec. Top. - Phys. Educ. Res.* **9** 1–22

[21] Begel A, Garcia D D and Wolfman S A 2004 Kinesthetic learning in the classroom *SIGCSE Bulletin (Association for Computing Machinery, Special Interest Group on Computer Science Education)* vol 36

[22] Garneli V, Giannakos M and Chorianopoulos K 2017 Serious games as a malleable learning medium: The effects of narrative, gameplay, and making on students' performance and attitudes *Br. J. Educ. Technol.* **48** 842–859

Journal of Physics: Conference Series **2126** (2021) 012016 doi:10.1088/1742-6596/2126/1/012016

Simple Harmonic Motion Electronic Teaching Materials Based on Authentic Learning to Train Students' Problem-Solving Skills: Aspects of Validity

S R Saputri[1], M Wati[1], M Misbah[1]

[1]Physics Education Study Program, Universitas Lambung Mangkurat, Indonesia

*shofiarihtazkias@gmail.com

Abstract. The absence of teaching materials that can train students' problem-solving skills optimally is the background of this research. This study describes the validity of simple harmonic motion electronic teaching materials based on authentic learning to train students' problem-solving skills. This research is development research with the ASSURE model that focuses on the Utilization of Instructional Materials stage. The instrument used is the validity sheet of electronic teaching materials. The validity of electronic teaching materials is viewed from the aspect of content, construct, and appearance. The validation of electronic teaching materials is carried out by three validators: two expert validators and one practitioner validator. The data analysis technique used descriptive statistics. The results showed that the validity of electronic teaching materials with an average score of 3.87 was categorized as good. It was concluded that the simple harmonic motion electronic teaching material based on authentic learning was declared valid to train students' problem-solving skills. Furthermore, these electronic teaching materials can be used as teaching materials in the classroom to determine the practicality and effectiveness of the developed teaching materials.

1. Introduction

Students in school need teaching materials that are closely related to the daily life of students so that students do not get bored in learning. Teaching materials that present real problems in physics learning are expected to train students' problem-solving skills. Problem-solving skills are generally defined as a person's ability to engage in cognitive processing to understand and solve a problem [1]. This problem-solving ability can be trained by suitable teaching materials, namely teaching materials related to physical matter and phenomena around [2]. So, such teaching materials are needed by teachers.

Based on the observations at one of the public high schools in Banjarmasin, it was obtained that students' problem-solving skills are still low. The lack of learning resources such as manuals is one of the contributing factors to learners' low level of problem-solving skills. In addition, the low mathematical ability of learners also affects the common problem-solving ability. Reddy & Panacharoensawad [3] also added that the low problem-solving skills of learners are caused by the inability of learners to understand the content of the material and the principles of physics in the form of formulas. Therefore, teachers should take initial steps to address the problem.

The solution that can be done to train students' problem-solving skills in physics subjects is to develop authentic learning-based teaching materials. Authentic learning in learning design is essential to create learning opportunities for students [4]. The use of authentic learning shows that knowledge and skills must be learned in a real-world context [5].

Authentic learning-based teaching materials are worth using to train students' problem-solving skills. Teaching materials that authentic associate learning with the region's culture can improve learners' understanding in practising problem-solving skills [6]. Using contextual-based teaching materials by integrating local wisdom is very used in physics learning [2]. Incorporating authentic learning in education allows long-term retention of what has been learned and improves understanding [7]. Therefore, students' problem-solving skills can be trained by developing simple harmonic motion electronic teaching materials based on authentic learning.

Integrating authentic learning-based teaching materials with the local wisdom of learners can improve learners' understanding in practising problem-solving skills. There are various physical concepts related to the local wisdom of South Kalimantan, one of which is the local wisdom of "Baayun Maulid", related to simple harmonic motion.

Researchers believe that developing authentic learning-based electronic teaching materials will pique learners' interests and make it easier for students to understand and solve the problems presented in the teaching materials. Bruner's theory of learning underlies learning with problem-solving, which explains that learners are encouraged to think and learn through problem-solving activities [8]. In addition, authentic learning must be set up so that students have the opportunity to practice their problem-solving skills [9]. Authentic learning is based on a constructivist view in which students create their understanding [10].

During pandemics like this, the need for teaching materials in electronic form is indispensable. One of the applications used to create electronic teaching materials is flip PDF professional. Teaching materials developed with professional PDF flip applications are worth using in the learning process [11] [12]. Flip PDF Professional has several advantages, including the resulting teaching materials that can be inserted into images, videos, animations, and simulations that can absorb interactive learning mediums so that learning becomes monotonous.

The use of electronic teaching materials can transform physics from less-liked subjects into students' favourite subjects. On the other hand, there is a significant change in the attitude of learners [13]. Electronic teaching materials are interactively designed and incorporate technologies that evolve [14]. The content of students' interactive electronic modules varies more by integrated inserting motion pictures (videos), animations, simulations, and materials.

The absence of authentic learning-based electronic teaching materials on this simple harmonic motion material caused researchers to produce the teaching material. Through this teaching material is expected to train students' problem-solving skills. This study aims to describe the validity of authentic learning-based electronic teaching materials using the Flip PDF Professional application. Electronic teaching materials are expected to be used as teaching materials that help students learn independently during the Covid-19 pandemic.

2. Methods

This study uses the ASSURE development model. This development model consists of four stages, as shown in Figure 1.

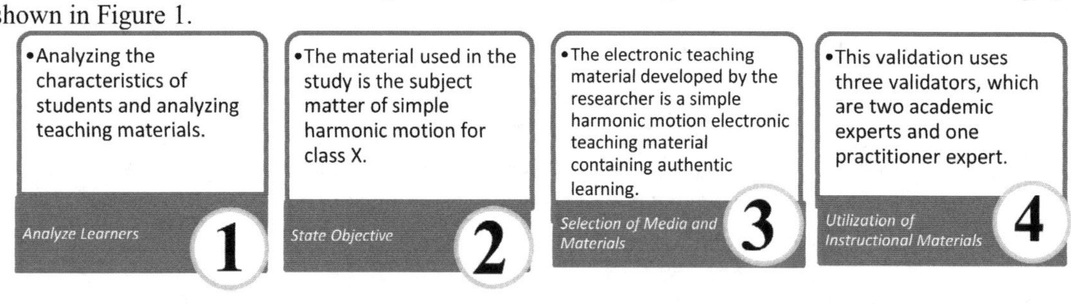

Figure 1. Development stages

The validity of electronic teaching materials is reviewed from the aspects of content, construct, and appearance. Data on the validity of teaching materials was obtained from the assessment of 3 validators, namely two expert validators and one practitioner validator. Electronic teaching material validation sheet instruments using scales 1 to 5. The validation result data is calculated in the form of an average score for each aspect of the assessment, then adjusted to the assessment criteria in Table 1.

Table 1. Criteria for validation of electronic teaching materials

No	Average score	Category
1	$\bar{x} > 4{,}2$	Very good
2	$3{,}4 < \bar{x} \leq 4{,}2$	Good
3	$2{,}6 < \bar{x} \leq 3{,}4$	Sufficiently good
4	$1{,}8 < \bar{x} \leq 2{,}6$	Less good
5	$\bar{x} \leq 1{,}8$	Not very good

3. Results and Discussions

The product developed is authentic learning-based electronic teaching materials adapted to the characteristics of simple harmonic motion materials and the characteristics of X-grade students at the high school level. Electronic teaching materials developed using multimedia software that is flip pdf professional. This electronic teaching material aims to train students' problem-solving skills and their application through direct teaching-learning models.

At the beginning of the meeting, authentic pictures and questions motivate students to be actively involved in learning. The electronic teaching materials developed contain authentic learning information about simple harmonic movements related to the environment of the learner's area, as stated in Figure 2 (a). Early knowledge in this electronic teaching material is presented in the corner of the recall (Figure 2b). As well as physical events related to learners' daily life, it is loaded in your know-how section. Electronic teaching materials also present an understanding test that is completed individually, as stated in Figure 2c.

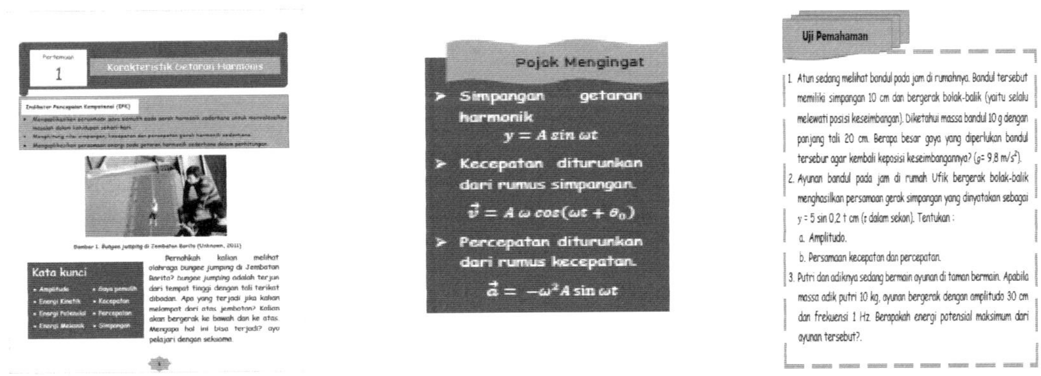

(a) (b) (c)

Figure 2. a) Initial meeting view; b) Remembering corner view; c) Understanding test display

Competency tests are also provided on electronic teaching materials that are developed to find out the level of students' understanding of the lessons they have learned. The final part of the teaching material contains a bibliography as a reference source information used by researchers in developing electronic teaching materials. Problem-solving in this electronic teaching material uses Heller's problem-solving stages. First of all, students will be required to visualize the problem; at this stage, students will describe the physical situation of the physics problem presented. Then students must describe the problem, namely writing down the quantities that are known and asked. Then students plan solutions; at this stage, students write down concepts, principles, equations, and mathematical stages in a coherent way to solve problems. Furthermore, students carry out the planned solution by substituting the values of the related physics quantities into mathematical stages. Finally, students evaluate solutions regarding the problem-solving stages that students have done. Three validators have validated this electronic teaching material. Here are the differences between the before and after validation teaching materials, as stated in Table 2.

Table 2. Differences before and after validation

Before	Validator Suggestion	After
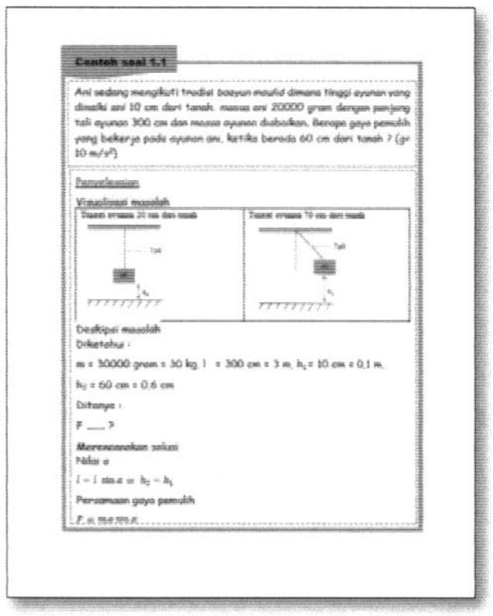	Every problem-solving activity be adapted to reality	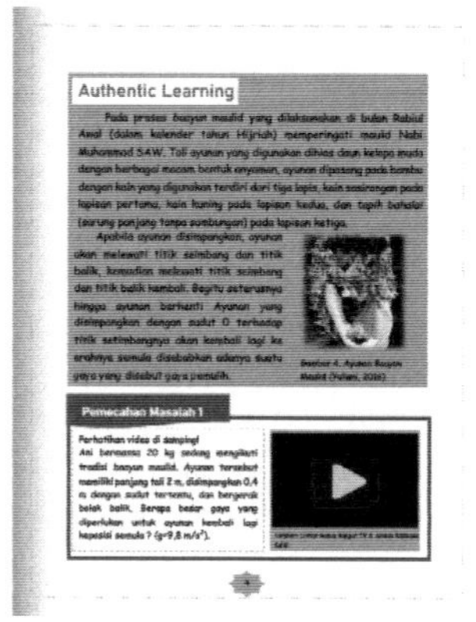
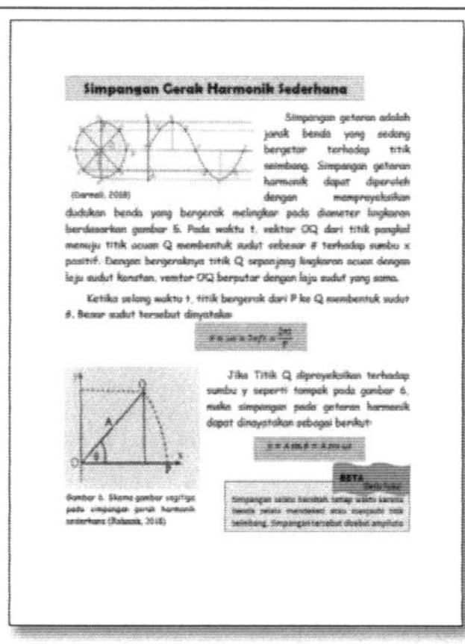	Add motivational content	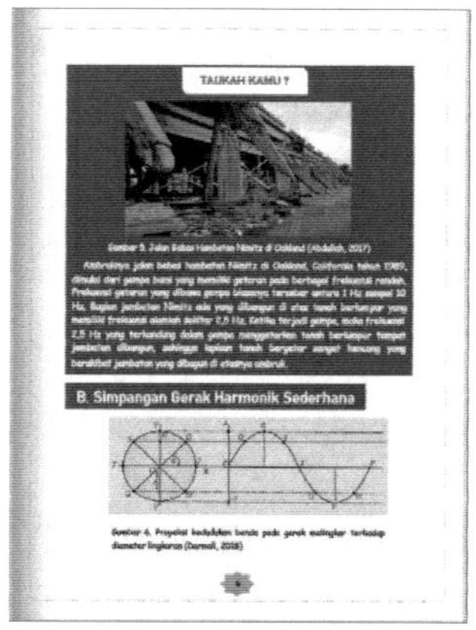
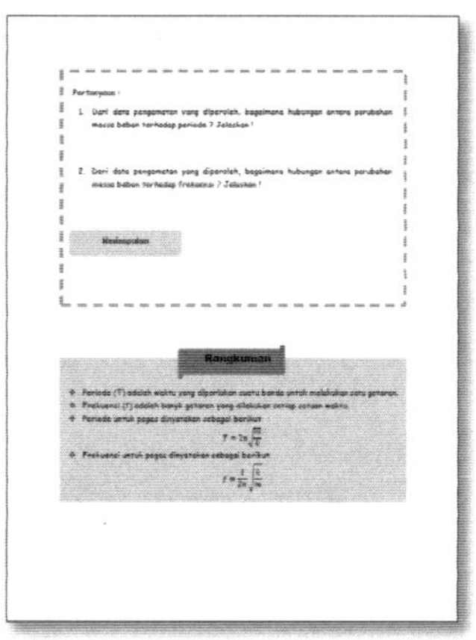	Work on student worksheets online	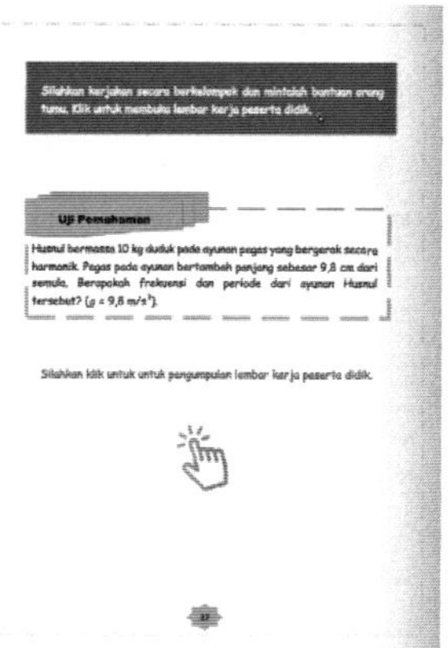

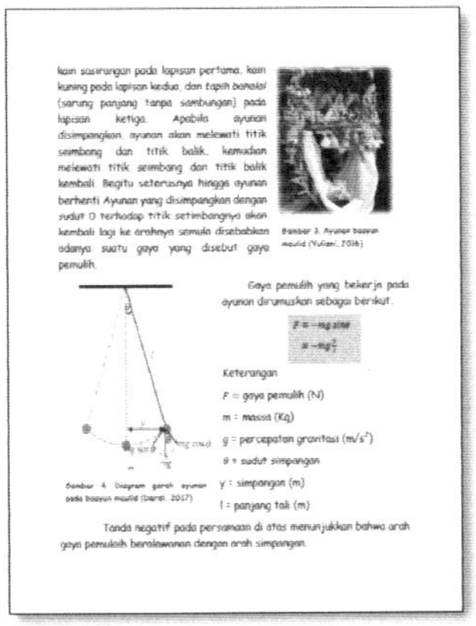	Explain authentic learning that is applied to teaching materials	

Based on Table 2, it can be seen some differences between electronic teaching materials before and after revision. Revisions are made following suggestions and input from the validator. The validator requests that every problem-solving activity be adapted to reality. Therefore the author includes an illustration video to strengthen the authentic learning content in this electronic teaching material. This allows students to observe directly through video to accelerate students' understanding in solving problems. The validator also asked to add motivational content to the teaching materials to increase students' enthusiasm in the learning process, so the author made a "Tahukah Kamu" presentation in the teaching materials to add insight and knowledge of students. The validator suggests that it is better to work on student worksheets online, and this can be done directly through teaching materials to shorten the time of learning implementation. In addition, the collection of student worksheets can be done quickly and in a short time to make the learning process more practical. To improve local wisdom in electronic teaching materials, the Validator asks to explain authentic learning that is applied to teaching materials. The validity of teaching materials seen in the assessment aspect can be observed in Table 3.

Table 3. Validation results of electronic teaching materials

No	Assessment Aspects	Average	Category
1.	Content	3.84	Very Good
2.	Construct	3.89	Very Good
3.	Display	3.87	Very Good

This electronic teaching material was developed to provide students with convenience in understanding physics concepts in simple harmonic motion materials and their application into daily life. Teaching materials designed to train students' problem-solving skills can also simultaneously train their high level of thinking skills through activities to explore surrounding problems to obtain a better conceptual understanding [15].

This electronic teaching material consists of front cover, steering cover, foreword, introduction, table of contents, instructions for using electronic teaching materials, concept map, competency achievement indicators, keywords, material description, authentic learning, problem-solving, "Taukah Kamu", remembering corner, know the character, comprehension test, summary, essential material, competency test, glossary, bibliography, and answer key.

This electronic teaching material was developed about the internet subscription-based publication model to be accessible [16]. In addition, the development of electronic teaching materials using Flip PDF Professional is one of the solutions in answering educational challenges in the era of industrial revolution 4.0 [17].

Sub-material is divided into three, namely: characteristics of harmonic vibration, period and frequency of the spring, and period and frequency of simple swing. These electronic teaching materials are arranged in a systematic, sequential, and orderly manner. Systematics in this electronic teaching material aims to make students' thinking activities run regularly. Thus, students can follow the flow of the material and easily understand the material presented [18].

Electronic teaching materials developed regarding language and readability aspects of electronic teaching materials must meet several standards, such as using sentences according to reasonable and correct Indonesian language rules. The sentence structure used must also be communicative and following the stage of language mastery of students. The same thing is also explained by Arsanti [19] that the language of suitable teaching materials must be communicative and easy to understand by students.

The last evaluation is used for the achievement of learners and their conformity to the standards that have been set. Evaluations are based on predetermined indicators and learning objectives. This is so that teachers can know the weaknesses or difficulties that students still face. This is in line with Tosuncuoglu's opinion that teachers should evaluate in order to classify students, make improvements, provide feedback, and develop appropriate teaching [20].

Furthermore, the validation of the constructs of simple harmonic motion electronic teaching materials that were developed fulfils the characteristics of authentic learning mentioned by [21] in (1) learning-centred on authentic tasks interesting for students. Authentic tasks in the form of solving real problems that are relevant to the lives of students; (2) students are involved in digging and investigating activities; (3) learning is interdisciplinary; (4) students learn to be active and driven by the students themselves, while teachers, parents, and resource persons are helpful or directing; (5) the teacher uses a scaffolding strategy, namely assisting as needed and letting students work freely when they can do it themselves; (6) students work with many sources (simple harmonic motion electronic teaching materials, and student worksheets); (7) students often work together and have broad opportunities to discuss to solve problems.

This electronic teaching material developed is designed using a direct teaching model. The direct teaching model helps learners learn declarative and procedural knowledge step by step. Through the direct teaching model, teachers can explain simple harmonic motion materials and problem-solving stages to students gradually. The problem-solving stage can be achieved well if the learning process in the classroom conditioned students are accustomed to solving the problem [22]. According to Arends [23], the direct teaching model is designed to gradually help learners master declarative knowledge and procedural skills. This reinforced the research results Amrita et al. [24] that direct teaching can improve students' problem-solving skills.

The developed electronic teaching materials are also made to train students in using their skills in solving a problem. Heller's problem-solving skills in this material are applied to solving sample questions and other practice questions in electronic teaching materials. Problem-solving skills [25], consist of 5 stages of problem-solving skills, namely (1) visualizing the problem; (2) describe the problem; (3) planning solutions; (4) implement solutions; (5) solution evaluation. This problem-solving stage is also the theoretical basis for developing simple harmonic motion electronic teaching materials containing authentic learning, which is contained in the evaluation criteria for the validation of electronic teaching materials.

Problem-solving skills are one of the essential skills of the 21st century, so it needs to be measured to know how readiness learners face the challenges of the 21st century [26]. This electronic teaching material presented authentic problems so that students are trained in solving problems in daily life. This authentic learning encourages learners to create a direct relationship between the new material being studied and their previous Mims [27]. Experiences like this will increase students' motivation.

Validation of the display of electronic teaching materials depends heavily on the role of visual communication. Layouts, colours, and themes on electronic teaching materials are made consistent at every meeting. The appearance of electronic teaching materials also relies heavily on the cover; the cover design is created with a combination of bright colours and an authentic cover image that attracts students to open and read the teaching material. Electronic teaching materials can be characterized as fully electronic hypertext interactive learning materials [28]. Thus, electronic teaching materials contain text and elements that can attract learners' attention.

The electronic teaching materials developed have a clear and appropriate numbering system, type, and size of letters, namely using the Comic Sans MS font with a font size of 14 which makes the appearance of the letters not monotonous and visible. In terms of design and space/layout in the content of electronic teaching materials, it is appropriate, with the balanced placement of authentic text and illustrations, to understand the material presented efficiently. According to [29], the presentation aspect relates to the method used in packaging teaching materials to be easy to understand and interesting for students. Judging from the presentation aspect, the teaching materials are arranged systematically and following the learning indicators. The teaching materials developed are also accompanied by language that invites or motivates. The goal is that students are more enthusiastic about completing all the stages contained in the teaching materials.

This electronic teaching material obtains a validity assessment with a good category. These results show that the teaching materials developed to meet the standards of good teaching material characteristics and can be used [30]. Thus the content of electronic teaching materials based on authentic learning can be tested in the learning process.

4. Conclusion

Simple harmonic motion electronic teaching materials based on authentic learning are declared valid and can train students' problem-solving skills. In addition, this teaching material can be used later to know the practicality and effectiveness of the developed teaching materials.

Acknowledge
Researchers would like to thank the Universitas Lambung Mangkurat for funding PNBP University in 2021, in the program “Dosen Wajib Meneliti”.

References

[1] Shute V J, Wang L, Greiff S, Zhao W and Moore G 2016 Measuring problem solving skills via stealth assessment in an engaging video game *Comput. Human Behav.* **63** 106–17
[2] Satriawan M and Rosmiati R 2017 Pengembangan Bahan Ajar Fisika Berbasis Kontekstual Dengan Mengintegrasikan Kearifan Lokal Untuk Meningkatkan Pemahaman Konsep Fisika Pada Mahasiswa *JPPS (Jurnal Penelit. Pendidik. Sains)* **6** 1212
[3] Reddy M V B and Panacharoensawad B 2017 Students Problem-Solving Difficulties and Implications in Physics : An Empirical Study on Influencing Factors *J. Educ. Pract.* **8** 59–62
[4] Herrington J, Reeves T C and Oliver R 2010 *A Guide to Authentic e-Learning* (London & New York: Murdoch University)
[5] Kartoglu U, Vesper J, Teräs H and Reeves T 2017 Experiential and authentic learning approaches in vaccine management *Vaccine* **35** 2243–51
[6] Ridho M H, Wati M, Misbah M and Mahtari S 2020 Validitas Bahan Ajar Gerak Melingkar Berbasis Authentic Learning Di Lingkungan Lahan Basah Untuk Melatih Keterampilan Pemecahan Masalah *J. Teach. Learn. Phys.* **5** 87–98
[7] Smeds P, Jeronen E and Kurppa S 2015 Farm education and the value of learning in an authentic learning environment *Int. J. Environ. Sci. Educ.* **10** 381–404
[8] Maulida M, Wati M and An'nur S 2015 Pengembangan Bahan Ajar Berbasis Pendidikan Karakter Dalam Mendukung Implementasi Kurikulum 2013 *Berk. Ilm. Pendidik. Fis.* **3** 11
[9] Iucu R B and Marin E 2014 Authentic Learning in Adult Education *Procedia - Soc. Behav. Sci.* **142** 410–5
[10] Roach K, Tilley E and Mitchell J 2018 How authentic does authentic learning have to be ? *High. Educ. Pedagog.* **2696** 1–15
[11] Sriwahyuni I, Risdianto E and Johan H 2019 Pengembangan Bahan Ajar Elektronik Menggunakan Flip Pdf Professional Pada Materi Alat-Alat Optik Di Sma *J. Kumparan Fis.* **2** 145–52
[12] Susanti T, Kurniadewi F and Nurjayadi M 2021 Development of Protein Metabolism

Electronic Module by Flip PDF Professional Application *J. Phys. Conf. Ser.* **1869**

[13] Zahorec J, Haskova A and Munk M 2010 Impact of Electronic Teaching Materials on Process of Education - Results of an Experiment *Informatics Educ.* **9** 261–81

[14] Suwatra W, Suyatna A and Rosidin U 2018 Development of Interactive E-Module for Global Warming to Grow of Critical Thinking Skills *Int. J. Adv. Eng. Manag. Sci.* **4** 543–9

[15] Trevathan J and Myres T 2013 Towards Online Delivery of Process - Oriented Guided Inquiry Learning Techniques in Information Technology Courses Motivation for adopting POGIL for use in online courses *IEEE Trans. Syst. Man, Cybern. Part ASystems Humans* **6** 1–11

[16] Sawyer S K 2002 Electronic books: Their definition, usage and role in libraries *Libres* **12**

[17] Misbah M, Sasmita F D, Dinata P A C, Deta U A and Muhammad N 2021 The validity of introduction to nuclear physics e-module as a teaching material during covid-19 pandemic *IOP Conf. Ser. Earth Environ. Sci.* **1796**

[18] Prasetiyo N A and Pertiwi P 2017 PENGEMBANGAN BUKU AJAR BERBASIS LINGKUNGAN HIDUP PADA MATAKULIAH BIOLOGI DI UNIVERSITAS TRIBHUWANA TUNGGADEWI The Development of Environment based Textbook in Biology Course at Tribhuwana Tunggadewi University *J. Pendidik. Biol. Indones.* **3** 19–27

[19] Arsanti M 2018 Pengembangan Bahan Ajar Mata Kuliah Penulisan Kreatif Bermuatan Nilai-Nilai Pendidikan Karakter Religius Bagi Mahasiswa Prodi Pbsi, Fkip, Unissula *KREDO J. Ilm. Bhs. dan Sastra* **1** 71–90

[20] Sasmita F D 2020 *Pengembangan modul elektronik bermuatan literasi sains melalui pengajaran langsung pada materi hukum gravitasi newton skripsi*

[21] Budiyanto M A K 2016 *Sintak 45 Model Pembelajaran dalam Student Centerd Learning (SCL)*

[22] Susanto H A 2015 *Pemahaman Pemecahan Masalah Berdasar Gaya Kognitif* vol 1 (Yogyakarta: Deepublish)

[23] Arends R I 2015 *Learning to Teach* (New York: The McGraw Hill Companies)

[24] Amrita P D, M. Arifuddin J and Misbah M 2016 MENINGKATKAN KEMAMPUAN PEMECAHAN MASALAH SISWA MELALUI MODEL PENGAJARAN LANGSUNG PADA PEMBELAJARAN FISIKA DI KELAS X MS 4 SMA NEGERI 2 BANJARMASIN *Berk. Ilm. Pendidik. Fis.* **4** 248–61

[25] Heller P, Keith R and Anderson S 1992 Teaching problem solving through cooperative grouping. Part 1: Group versus individual problem solving *Am. J. Phys.* **60** 627–36

[26] Hidayat S R, Setyadin A H, Hermawan H, Kaniawati I, Suhendi E, Siahaan P and Samsudin A 2017 Pengembangan Instrumen Tes Keterampilan Pemecahan Masalah pada Materi Getaran, Gelombang, dan Bunyi *J. Penelit. Pengemb. Pendidik. Fis.* **3** 157–66

[27] Mims C 2003 Authentic learning: A practical introduction & guide for implementation *Meridian* **6**

[28] Klement M, Dostál J and Marešová H 2014 Elements of Electronic Teaching Materials with Respect to Student's Cognitive Learning Styles *Procedia - Soc. Behav. Sci.* **112** 437–46

[29] Habibi M, Chandra C, Mahyuddin R and Hendri S 2018 Validity of Teaching Materials for Writing Poetry Based on Creative Techniques in Elementary Schools *Mimb. Sekol. Dasar* **5** 145

[30] Sugiyono 2015 *Metode Penelitian Pendidikan. Pendekatan Kuantitatif, Kualitatif, dan R&D* (Bandung: Alfabeta)

Journal of Physics: Conference Series **2126** (2021) 012017 doi:10.1088/1742-6596/2126/1/012017

The influence of science, technology, engineering, and mathematics (STEM) learning approaches on learning outcomes

P Andriani[1*], Muhammad Ali[1], and M Jarnawi[1]

[1]Sub-Department of Physics Education, Faculty of Teacher Training and Education Universitas Tadulako, Jl. Soekarno Hatta KM. 9 Kampus Bumi Tadulako Tondo Palu-Central Sulawesi, Indonesia.

*Corresponding author: dwi3.ranie@gmail.com

Abstract. The objective of the study is to discover the effect of the Science, Technology, Engineering, and Mathematics (STEM) learning approach on the physics learning outcomes The study is considered as quasi-experimental research by implementing the non-equivalent, pretest-posttest as the research design. There were two science classes of SMA Negeri 1 Sindue Tobata were selected as the sample of the study. The Science, Technology, Engineering, and Mathematics learning approach was applied to the experimental class while the scientific approach was applied to the control class. The learning outcomes test obtained showed that the average score of the experimental group was 15.95 and the average score of the control group was 12.25. Based on the results of hypothesis testing, obtained that t_{count}=3.00 and t_(0.0975(38))=2.024 at a significance level of a=0.05. It means that the value of t_count is outside the acceptance area of H_0. The findings concluded that there were differences in physics learning outcomes between the students who were taught by using STEM approach and the students who were only taught by implementing the scientific approach in SMA Negeri 1 Sindue Tobata.

1. Introduction

The industrial revolution of education in the 21st century is the education that requires each student to have skills, knowledge, and abilities in the fields of technology, media and information, learning skills, innovation, life skills, and competencies [1–3]. The 21st century is well known as the century of knowledge, a century in which science and technology are developing rapidly. Therefore, the resources that have high competitiveness in the globalization era are highly needed. The technological developments are not possible to be separated from the contribution of physical science development since physics is a science that studies natural events and phenomena [4]. Physics is one of the interesting studies because it is directly correlated to the actual phenomenon and can also be applied in daily activities.

Lesson delivering process in physics learning should involve the concepts and real life phenomena [5], because the comprehension in physics requires more than just memorizing skills [6]. Teachers should be able to create learning environment that could enable students to think, assume and construct the knowledge independently [7]. However, the physics lesson in reality tends to be taught with the concept of memorizing. It causes many students to experience difficulties in learning physics especially when they implement the concepts of physics in their daily life [8]. It occurs because teachers usually give less relevant examples of physics subjects with real life phenomena.

The emphasis on comprehending the basic concepts and the basic understanding of science is not in accordance with matters relating to everyday life, whereas Harefa [9] reveals that physics is a science that is closely related to human life. It is what causes the low ability of students either in the field of physics or other fields of studies, which then also causes low student learning outcomes.

Based on the study conducted by Fatmawati [10] at SMA 4 Jember, it can be known that the low students' learning outcomes are caused due to the tendency of the teacher to apply the conventional learning method in the educational process, in which the process of learning only centered to the teacher while the students can only listen to the material explained by the teacher. It causes students to be less active in the learning process as well as the tendency of the students to only memorize the material but do not understand the topic being taught. In addition, since the educational process is central to the teachers, they tend to only elaborate the material monotonously without correlating the concepts of physics to everyday life. It is actually causing the low students' learning outcomes. Therefore, the learning approach which able to connect the physics subject into everyday activity is needed so that the learning process could be more meaningful and improve students' activeness in the learning process so that the students not only memorize but also comprehend the material being taught. The solution to overcome the problem is to apply an approach in the learning process, one of which is the STEM approach.

The STEM approach is an approach that refers to four components of knowledge, namely science, technology, engineering, and mathematics [11]. In line with the statement, research shows that the application of STEM is able to facilitate the students to develop knowledge, help answer questions based on investigation, and can help students to create new knowledge [12].

STEM is designed to improve people's abilities in knowledge and innovative technology products in order to be able to compete globally [13]. The aim of STEM education for students is expected to be able to deliver students to meet 21st-century abilities, including learning and innovation skills. The STEM learning approach by integrating the four components is able to improve students' physics learning outcomes which are characterized by the ability in solving problems, making decisions, analyzing assumptions, evaluating and conducting investigations. The STEM approach which is integrated in learning has the potential to equip students with 21st-century work skills through learning outcomes [14]. Learning outcomes are an indication of the ability of students' work skills, the extent to which the contribution of the influence of STEM learning on learning outcomes in the form of creativity, problem-solving skills, and students' self-motivation.

The implementation of the STEM approach is very important to student learning outcomes, it is supported by the research conducted by Muthi'ik, it was concluded that the STEM learning approach is effective for increasing student self-efficacy and student learning outcomes [15].

According to the research conducted by [16], it was concluded that STEM learning trains students' creative abilities in connecting the four fields of exact disciplines so that students have deep and dynamic insights in solving global issues. Then STEM learning also affects student learning outcomes. The purpose of this study is to determine whether there is an influence of the Science, Technology, Engineering, and Mathematics learning approach on the physics learning outcomes of the students of SMA Negeri 1 Sindue Tobata. The advantages of this study are as consideration for selecting the learning model to be applied to achieve optimal physics learning outcomes

2. Methods

The study was quasi-experiment research, with the non-equivalent pretest-posttest control group design. The treatment group was taught using science, technology, engineering, and mathematics (STEM) approach, while the control group was taught using scientific approach. The samples were 11th graders of science classes and were selected using the purposive sampling. The two selected classes are classes that are considered academically homogeneous. Based on the observation, the number of students in both classes was equal and have homogeneous learning outcomes, which is seen from the learning outcomes at the previous semester in the final semester examination in which the average score obtained by the students in both classes was 86.

The technique of data collection in this study was in the form of object tests and observation sheets. Objective tests were used to measure student learning outcomes after learning treatment was given. While the observation sheets were used to assess the activities of students and teachers during the learning process.

The research instrument used was an objective test for multiple-choice developed by the researcher. This test has been checked by expert validators. This special test for learning outcomes consists of 25 numbers that were used twice, such as at the beginning (pretest) and the end (posttest) which were in accordance with the indicators. After that, the instrument was tested to measure the validity, the level of difficulty of the questions, the distinguishing power, and the reliability of the items. The instrument used in this study was first checked by the validator, before being tested and used for research. The results of the statistical test obtained that 21 items were accepted, 4 were revised and 25 items were rejected because they did not meet the criteria.

The normality, homogeneity, and hypothesis testing were carried out to analyze the result of the study. Normality testing was carried out to see whether the data obtained from the research results were normally distributed or not, the homogeneity of variance test was used to determine whether the variance between the two classes was the same or different, and hypothesis testing was used to see whether the formulated hypothesis was supported by the data that has been collected.

3. Results and Discussion

The pretest-posttest of students' learning outcomes can be seen in Table 1. Overall, the posttest results increased for both the control and treatment groups.

Table 1. Results of data analysis of pretest and posttest of student learning outcomes

Description	Pretest		Posttest	
	Experimental	Control	Experimental	Control
Sample (n)	18	18	18	18
Minimum score	8	7	14	8
Ideal score	30	30	30	30
Maximum score	16	16	27	23
Average score	11.28	11.39	19.50	16.94
Standard dev.	2.38	2.68	3.41	4.45

To test the hypothesis, normality and homogeneity test were conducted. The results of these tests are presented in Table 2 and 3.

Table 2. Normality of the final test distribution at the experimental and control class

Description	Posttest	
	Experimental	Control
Sample	20	20
χ^2_{count}	2.29	2.58
χ^2_{table}	5.99	5.99

According to the Table 2 by using the Chi-Square test with acceptance criteria $x^2_{count} < x^2_{table}$ in which the final test of x^2_{count} value obtained is smaller than the value of x^2_{table}. Based on the decision-making criteria, H_1 is accepted. These results indicate that the data from the experimental class and control class are normally distributed. Meanwhile, information in Table 3 indicated that the value of F_{count} is lower than the value of F_{table}. In the acceptance criteria, it can be concluded that the data in the experimental class and control class have equal variance (homogeneous).

Tabel 3. The result of homogeneity post-test of experimental class and control class

Description	Posttest	
	Experimental	Control
Variance Score	15.4	15.42
F_{count}	1.03	
F_{table}	2.17	
Criteria	Homogeneous	

Hypothesis test was conducted using the parametric-based on the data. The detailed results can be seen in Table 4.

Table 4. The t-test on the final test (posttest)

Description	Final Test (posttest)
t_{count}	3.00
$t_{table}(\alpha = 0.05)$	2.024
Criteria	H_1 accepted

Based on the data in Table 4, the $t_{count} > t_{table}$ or 3.00 > 2.024. It shows that the value of t_{count} is in the rejection area of H_0, thus the H_1 is accepted. IT denotes that there is an effect of the implementation of STEM approach on students' learning outcomes on physics. The results indicate that students' cognitive achievement in treatment were better compared to students who were in control group. It is owing to the implementation of STEM in the teaching and learning process, which enables students to examine the lesson learned that occur in daily life. In the learning process conducted with the STEM approach, the teacher guided students to be able to observe daily-life problems (*Observe*) that were displayed by using video assistance (*Technology*). The problems were associated with physics (*Science*), which was the concept of sound waves. After that, brainstorming was conducted to stimulate thoughts and new ideas that can be used to solve the problems. The teacher could provide explanations about technology engineering techniques (*Engineering*) by utilizing the concept of sound waves. The teacher afterward could give guidance to the students to be innovative (*Innovation*) in developing the ideas that have been obtained by the students through group discussion along with their friends. The teacher could also guide the students to be able to model a physics problem into the form of a mathematical formula (*Mathematics*). Subsequently, the teacher guides students to be able to think creatively (*Creativity*) in solving physics problems using the assistance of LKPD. Finally, the last process carried out by the teacher in learning is directing students to be able to draw conclusions and relate the material to everyday life (*Society*).

Meanwhile, the scientific approach was implemented for the learning process in the control class. In contrast to the STEM approach which requires students to be able to think innovatively and creatively in solving problems, the scientific approach requires students to solve problems through careful planning activities, data collection, data analysis, and drawing conclusions. Problem solving steps involve observing, questioning, gathering information, reasoning, and communicating. Students were provided worksheets after lesson delivering. During problem solving processes, students could question the teacher and find the solution and required information through peer discussion. At the reasoning step, the teacher asks each group to summarize the results of the discussion and draw conclusions. Finally, at the last step to communicate the teacher assigns representatives of each group to present the results of the discussion. However, the application of the scientific approach in the learning process was still less effective. It was due to the uneven ability of students. Students were not familiar with the stages, especially during scrutinizing the correct information among the abundant information in the Internet. In addition, there were many errors done by students during the experiments which resulted in inference errors [17]. Although the students in STEM group outperformance the scientific group, both groups indicated an increase in learning outcome. It denoted that the scientific approach also give positive impacts on learning [18–20].

4. Conclusion

Based on the result, it could be concluded that there is an effect of learning outcome of students who were taught using the STEM approach in physics. However, despite the positive impact on physics learning, STEM application requires a longer time in the learning process. On that account, good preparation is necessary to be apply to integrate this approach in teaching physics.

Acknowledgement

We highly acknowledge the support of the Program Kompetisi Kampus Merdeka (PK-KM) Grant of the Education, Culture, Research, and Technology of the Ministry of the Republic of Indonesia, with funding identity: 119/E1/KM.05.03/2021.

References

[1] Wijaya E Y, Sudjimat D A and Nyoto A 2016 Transformasi pendidikan abad 21 sebagai tuntutan pengembangan sumber daya manusia di era global *Pros. Semin. Nas. Pendidik. Mat. 2016* **1** 263–78

[2] Boholano H 2017 Smart social networking: 21st Century teaching and learning skills *Res. Pedagog.* **7** 21-29

[3] Kivunja C 2015 Exploring the Pedagogical Meaning and Implications of the 4Cs "Super Skills" for the 21st Century through Bruner's 5E Lenses of Knowledge Construction to Improve Pedagogies of the New Learning Paradigm *Creat. Educ.* **06** 224–39

[4] D. C. Giancoli 2008 *Fisika* (Jakarta: Erlangga)

[5] Suriawati S and Mundilarto M 2019 SETS approach-based audiovisual media for improving the students' critical thinking skills *Psychol. Eval. Technol. Educ. Res.* **1** 95–103

[6] Samudra G, Suastra M and Suma M 2014 Permasalahan-Permasalahan Yang Dihadapi Siswa SMA Di Kota Singaraja Dalam Mempelajari Fisika *J. Pendidik. dan Pembelajaran IPA Indones.* **4**

[7] Nggadas D E P and Ariswan A 2019 The mastery of physics concepts between students are learning by ICT and laboratory experiments based-teaching *Momentum Phys. Educ. J.* **3** 21–31

[8] Surapranata S 2004 Peningkatan Pendidikan MIPA dalam Master Plan Pendidikan Indonesia *Booldet Seminar Nasional Penelitian Pendidikan dan Penerapan MIPA* (Yogyakarta)

[9] Harefa A R 2019 Peran Ilmu Fisika Dalam Kehidupan Sehari-Hari *War. Ed. 60*

[10] Fatmawati L, Sukidin S and Suyadi B 2018 Perbedaan Hasil Belajar Siswa Melalui Pendekatan Konvensional Dengan Pendekatan Saintifik Pada Kompetensi Dasar Fungsi Manajemen Siswa Kelas X Di Sma Negeri 4 Jember Tahun Ajaran 2016/2017 *J. Pendidik. Ekon. J. Ilm. Ilmu Pendidikan, Ilmu Ekon. dan Ilmu Sos.* **11** 134–9

[11] Mulyani T 2019 Pendekatan Pembelajaran STEM untuk menghadapi Rovolusi Industry 4.0 *Seminar Nasional Pascasarjana* (Semarang)

[12] Utami I S, Septiyanto R F, Wibowo F C and Suryana A 2017 Pengembangan STEM-A (Science, Technology, Engineering, Mathematic and Animation) Berbasis Kearifan Lokal dalam Pembelajaran Fisika *J. Ilm. Pendidik. Fis. Al-Biruni* **6** 67–73

[13] Winarni J, Zubaidah S and H S K 2016 STEM: apa, mengapa, dan bagaimana *Pros. Semin. Nas. Pendidik. IPA Pascasarj. UM* **1** 976–84

[14] Cachaper C, Spielman L J, Dahl Soendergaard B, Corwin S, Dietrich C, Rosenzweig M, Tabor L, Edmister W and Fortune J 2008 , "Universities as Catalysts for Community Building among Informal STEM educators: The Story of POISED", , New

York, *American Educational Research ssociation Conference* (New York)
[15] Muthi'ik 2018 *Efektivitas Penerapan Pendekatan Pembelajaran Stem Terhadap Self Efficacy Dan Hasil Belajar Siswa Menggunakan Model Pembelajaran Problem Solving Pada Materi Hukum Newton* (Lampung: Universitas Lampung)
[16] Muharomah D R 2017 *Pengaruh Pembelajaran STEM (Science, Technology, Engineering and Mathematics) Terhadap Hasil Belajar Pesrta Didik Pada Konsep Evolusi* (Universitas Islam Negeri Syarif Hidayatullah)
[17] Hosnan 2014 *Pendekatan Saintifik dan Kontekstual dalam Pembelajaran Abad 21* (Bogor: Ghalia Indonesia)
[18] Rahayu M S I and Kuswanto H 2020 Development of android-based comics integrated with scientific approach in physics learning *Journal of Physics: Conference Series* vol 1440
[19] Nurfathanah Amsal A E S and Arafah K 2020 Implementation of Scientific Approach in Physics Learning in Makassar City *J. Pendidik. Fis.* **8**
[20] Kaleka M and Nur F D M 2018 Experimental-Based Scientific Approach toward the improvement of Science Process Skill and Scientific Attitudes of Grade X Student MAN Ende *J. Sci. Educ. Res.* **2** 13–20

A Qualitative Study: Physics Concepts Used by Survivors in the 2018 Tsunami in Palu City

W A Suleman[1*], and U Wahyono[1]

[1]Sub-Department of Physics Education, Faculty of Teacher Training and Education Universitas Tadulako, Jl. Soekarno Hatta KM. 9 Kampus Bumi Tadulako Tondo Palu-Central Sulawesi, Indonesia.

*Corresponding author: wa44032@gmail.com

Abstract. This study aimed to identify and examine the Physics concepts used by victims of the tsunami disaster in Palu City based on the earthquake, tsunami and liquefaction events that occurred on September 28, 2018 in Palu City. This research was descriptive qualitative. Data were collected by observation, interviews and documentation. Based on the results of the study, the researchers found several Physics concepts that were applied when the victims made efforts to save themselves from the tsunami. These findings were discussed and grouped according to the theme of the Physics concept, including: the concept of speed and velocity, Newton's Second Law Concept, the concept of floating objects, and the concept of waves. The results can be used as a contextual example in learning physics. By doing so, the students will not only learn about Physics concepts, but also improve their understanding and awareness of local disasters.

1. Introduction

One of the major active faults in Central Sulawesi is the Palu-Koro fault, which according to Tjia [1], has a length of 300 km. The fault is capable of generating the greatest seismic risk and earthquake [2]. Previous studies indicated the risk of frequent earthquakes by the fast slip rate of the Palu-Koro Fault [3]. The estimation of slip rate of this fault is around 35 ± 8 mm year−1 [4]; 38 ± 8 mm year−1 [5]; and 40 mm year−1 [6]. These numbers indicate how this fault is highly active. Before the 2018 disaster, seven destructive earthquakes had occurred in Central Sulawesi within a decade, in 1905, 1907, 1909, 1927, 1937, 1968, and 2012 [7]. The potential of other earthquakes in this area is inevitable.

As Central Sulawesi is highly prone to earthquakes, tsunami hazard is also inevitable. For instance, when the hazardous disaster that hit Palu City in September 2018 for example, the tsunami occurred in less than 5 minutes after the major earthquake. It was a short-range tsunami or a local tsunami based on the time span between the main shaking and the tsunami which has a very fast arrival time of between 0-30 minutes.

However, most of the locals have a low risk perception associated with the tsunami. It is owing to the frequent minor earthquakes that have happened in Palu which were not followed by tsunamis. Accordingly, they believed that the area of Palu Bay is safe from this hazard. This low awareness of disaster risk perception is unfortunate since Central Sulawesi has the highest frequency of tsunami occurrence in Indonesia. In Palu City alone, before the accident in 2018, a tsunami that was triggered by the Palu-Koro fault, had happened four times in 1927, 1968 and 1996. Their arrival time was also less than 10 minutes. However, they were not well recorded. Documentation of the past natural events is significant for both the disaster-related studies and disaster-awareness dissemination activities. Fail to document the past events will affect the continuity of information dissemination.

In addition, disaster experiences that do not have a significant impact on victims can lead to reduced awareness of disaster risk in individuals [8]. For example, people fled to high places when a major earthquake occurred in 2005. Fortunately, it was not followed by a tsunami. However, it could drive individuals to believe that the future disaster will not be a big risk, or they will be able to overcome the risk caused by the disasters [9]. The different experiences while confronting natural hazards might cause different disaster risk perception, awareness and preparedness among locals regarding the potential occurrence of a tsunami. Therefore, it is imperative to sustain the information about the past events thus it can be used to raise awareness for the next generation about the disaster potential in their areas.

Based on the active movement of the Palu-koro fault, the potential for a tsunami in Palu Bay in the future is very high [10]. Therefore, this study aimed to record the stories of the tsunami's survivors and relate their stories to physics concepts. The results of the research are expected to be archived in the form of data and documentation of the tsunami history in Palu City. In addition, the stories of the survivors can be utilized in learning about disaster awareness since bringing contextual and familiar cases to students should be used as the key in disaster education [11].

2. Methods

This study was descriptive qualitative research, which examined facts about the physics concepts used by several survivors while escaping from the tsunami disaster in Palu City. Data divided into primary data, which were collected through observation and interview techniques, and secondary data through literature study and documentation related to the tsunami in Palu City. The selection of respondents used purposive sampling, especially those who have had direct experience during the 2018 tsunami. Some respondents were obtained from social media posts which were then offered for research interviews. The were 5 respondents (with various age ranges of 20 to 60 years), 2 men and 3 women. The data from the interviews were analyzed through three stages, namely data reduction, data presentation and conclusion drawing. The data were in the form of stories and were analysed from the perspective of physics concepts.

3. Results and Discussion

3.1 Speed and Velocity Concept

Speed and velocity concepts were based on the path taken by the survivors. The paths are illustrated as Figure 1. Kesya (17 years old) ran through the highway that was not perpendicular to the direction of the waves, and was hit by the waves. Meanwhile, Kesya's sister was not hit by the waves because she run through the old building which was perpendicular with the direction of the wave. From the story, the concept of velocity and speed applied. In terms of physics concepts with the same mileage, Kesya and Kesya's sister get different displacements. It can be said that with the same velocity, Kesya had smaller velocity than the direction of the tsunami wave, while Kesya's sister had a higher speed than the direction of the tsunami wave.

Journal of Physics: Conference Series **2126** (2021) 012018 doi:10.1088/1742-6596/2126/1/012018

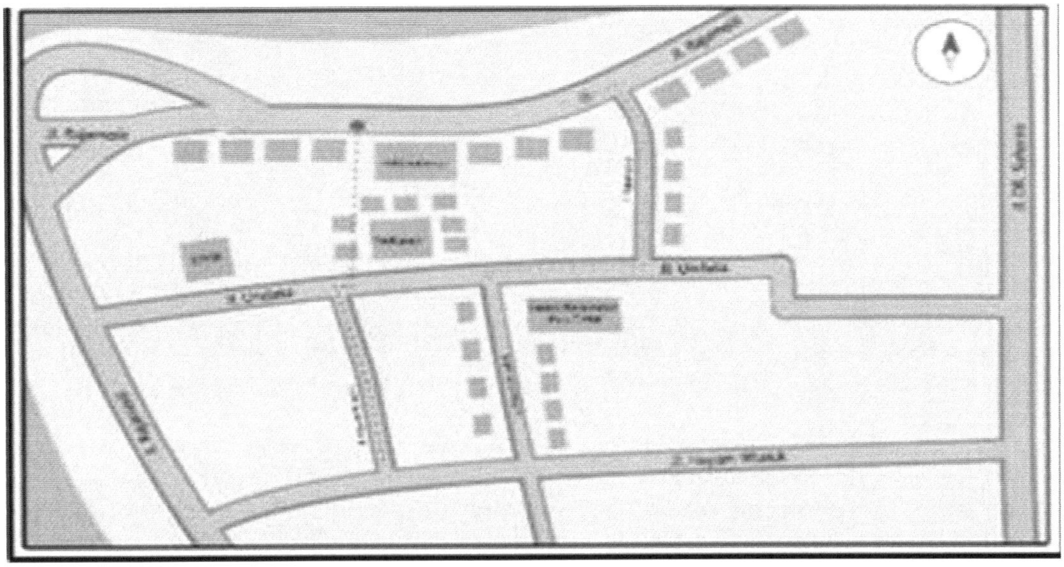

Kesya' route during tsunami
Kesya's brother route during tsunami

Figure 1. An overview of the paths taken by survivors

3.2 Floating Object Concept

Survivors used various efforts to save themselves from the waves. Some survivors managed to save themselves using trash cans to stay on the surface of the water. For example, a survivor stated that he was drowned multiple times by the waves before he grabbed a trash can. It helped him get an upward force or a buoyant force. In this situation, the survivors could stay afloat on the surface of the water without swimming or performing any movement that could drain their energy to stay on the water surface while trapped in the tsunami waves, at least until the tsunami waves recede or until help from the rescue team arrives.

3.3 Newton II Law

A survivor explained that she took off her shoes to get a wider range of motion and run faster. It applies to the concept of physics as stated in Newton's second law, where the acceleration of an object is directly proportional to the force imposed on it and inversely proportional to its mass. Conceptually, removing the shoes worn by the survivors is the same as reducing their mass. By reducing its mass, the force required to move or to run is becoming smaller. Then with this smaller mass, the survivor can increase the force and acceleration. Especially when they were hit and rolled by the waves, the survivors got more space and it made it easier for the survivors to swim up to the surface.

3.4 Wave

An amateur video about a ship caught in the tsunami waves indicated that the captain tried to head towards the middle of the sea or away from the coastal area. It implied that the concept of physics applied to this event is the concept of waves. As seen in the incident recorded in the video, the ship that will be leaning against the pier changes course away from the pier or the coast because they understand that on the coast, the tsunami waves become higher than in the middle of the sea. By directing the ship away from the pier, the height of the waves hitting the ship will be lower.

In Physics, tsunami waves can be explained from the phenomenon of transverse propagation of waves; its energy is a function of its height (amplitude) and its velocity. Its height is greatly influenced by its wavelength (see the comparison of depth, velocity and wavelength in Figure 2). Tsunamis have wavelengths of hundreds of kilometers, but in deep oceans behave like shallow-water waves. As a result, people who are on a ship sailing in deep sea or ocean will not be aware of a tsunami.

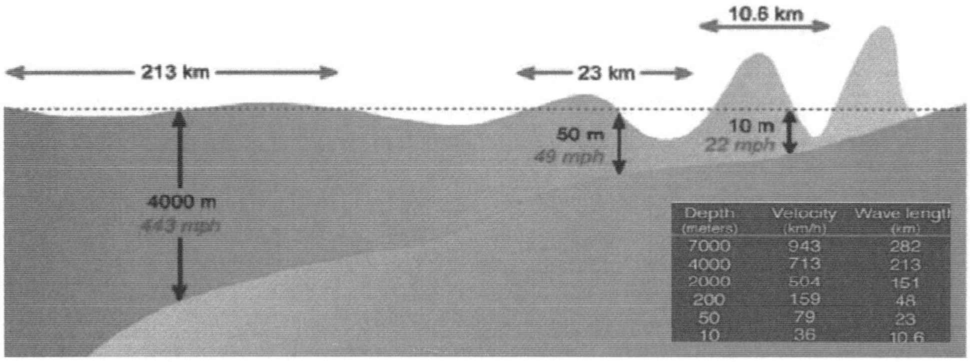

Figure 2. The comparison of depth, velocity and wavelength when it reaches the mainland.

The closer you get to the mainland or when you are at a shallower sea depth, the wavelength will decrease. However, the energy present in the waves will not decrease much. This is in accordance with the relationship between the rate of energy loss (energy loss rate) in the traveling wave is inversely proportional to the wavelength. It depicts that the larger the wavelength, the less energy will be wasted and the energy in a tsunami wave can be considered constant. Since the energy is constant, decreasing the speed will make the wave height (amplitude) increase.

Based on the descriptions of the results, there are some physics concepts used by survivors to save themselves. Referring to the opinion of Saleh and Pendley [11] who suggested that the inclusion of disaster cases should be a center for learning about disasters, incorporating victims' stories in Physics concepts as previously mentioned is reasonable. It is in line with Rae's statement [12] that an accident or a disaster experience should be a part of the core of safety curricula. Moreover, since the study of Wulandari et al [13] found that the linkages between tsunami-related topics in textbooks used by schools were not very relatable to the real-life situation, incorporating a real-life tsunami situation could be fruitful for students' better understanding about the events. Further, including disaster experience cases into Physics concepts will benefit in some ways. First, contextual cases increase student engagement because it is able to connect real-world events with what is being learned, and improve the relationship between what students know and what they will learn [14,15]. Second, while studying Physics, students also learn about disaster risk and safety.

4. Conclusion

Based on the analysis of some events and behaviors of survivors during survival attempts from tsunami, some can be related to physics concepts, consisting of the concept of speed and velocity, Newton's Second Law, floating objects and waves. The results of this study can be used as the linkage between physics, disaster and real-life situations. Despite the positive suggestion about the real-life situation inclusion into learning process, further investigation on how the influence of real-life hazardous experience on both the physics and safety and disaster learning is necessary.

Acknowledgement

We highly acknowledge the support of the Program Kompetisi Kampus Merdeka (PK-KM) Grant of the Education, Culture, Research, and Technology of the Ministry of the Republic of Indonesia, with funding identity: 119/E1/KM.05.03/2021.

Reference

[1] Tjia H D 1978 Active faults in Indonesia *Bull. Geol. Soc. Malaysia* **10** 73–92

[2] Watkinson I M and Hall R 2017 Fault systems of the eastern Indonesian triple junction: Evaluation of Quaternary activity and implications for seismic hazards *Geol. Soc. Spec. Publ.* **441** 71–120

[3] Patria A and Putra P S 2020 Development of the Palu–Koro Fault in NW Palu Valley,

Indonesia *Geosci. Lett.* **7** 1–11
[4] Bellier O, Siame L, Beaudouin T, Villeneuve M and Braucher R 2001 High slip rate for a low seismicity along the Palu-Koro active fault in Central Sulawesi (Indonesia) *Terra Nov.* **13** 463–70
[5] Stevens C, Mccaffrey R, Subarya C and Vigny C 1999 Rapid rotations about a vertical axis in a collisional setting revealed by the Palu fault , Sulawesi , Indonesia **26** 2677–80
[6] Socquet A, Hollingsworth J, Pathier E and Bouchon M 2019 Evidence of supershear during the 2018 magnitude 7.5 Palu earthquake from space geodesy *Nat. Geosci.* **12** 192–9
[7] Daryono 2018 Sejarah Kegempaan di Sesar Palu-Koro' in Tim Pusat Studi Gempa Nasional *Kajian Gempa Palu Provinsi Sulawesi Tengah 28 September 2018 (M7.4)* ed T P S G National pp 1–332
[8] Paton D, Johnston D, Mamula-seadon L and Kenney C M 2014 Recovery and Development: Perspectives from New Zealand and Australia *Disaster and Development:Examining Global Issues and Cases* (New York: Springer)
[9] Becker J S, Paton D, Johnston D M, Ronan K R and McClure J 2017 The role of prior experience in informing and motivating earthquake preparedness *Int. J. Disaster Risk Reduct.* **22** 179–93
[10] Sarapang, H, T., Rogi, O., Hanny P 2019 Analisis Kerentanan Bencana Tsunami di Kota Palu *J. Spasial* **6** 432–9
[11] Saleh J H and Pendley C C 2012 From learning from accidents to teaching about accident causation and prevention: Multidisciplinary education and safety literacy for all engineering students *Reliab. Eng. Syst. Saf.* **99** 105–13
[12] Rae A 2016 Tales of disaster: the role of accident storytelling in safety teaching *Cogn. Technol. Work* **18** 1–10
[13] Wulandari A T, Fauzi A, Darvina Y and Syafriani 2019 Analisis Kesesuaian Materi Fisika untuk Diintegrasikan dengan Materi Tsunami pada Buku Teks Pelajaran Fisika SMA/MA *J. Pillar Pgysics Educ.* **12** 449–56
[14] Sears S 2003 *Introduction to Contextual Teaching and Learning* (Bloomington: Phi Delta Kappa Educational Foundation)
[15] Hudson C C and Whisler V R 2013 Contextual teaching and learning for practitioners *Syst. Cybern. Informatics* **6** 54–58

Development of nano physics learning media (physics monopoly game) based on software

M Hendri[1] and M Jarnawi[1*]

[1]Sub-Department of Physics Education, Faculty of Teacher Training and Education Universitas Tadulako, Jl. Soekarno Hatta KM. 9 Kampus Bumi Tadulako Tondo Palu-Central Sulawesi, Indonesia.

*Corresponding author: i.am.jarnawi@gmail.com

Abstract. This study aims to develop NaNo Physics learning media (Physics Monopoly Game) on work and energy topic. The type this research was research and development (R&D) with the ADDIE model. Data were analysed using the descriptive statistical analysis. The results indicated that the physics monopoly was feasible to be used as a learning media. The overall results of the assessments of material and media experts, and students from individual trials, small group trials and field trials showed that they had met the eligibility requirements of the developed physics monopoly learning media. In addition, according to the field testing, the developed product can increase students' learning motivation from an average score of 3.19 which was in the "enough" category and increased to 4.39 in the "very high" category.

1. Introduction

Learning activities are complex. It is a system that is composed of components that are interrelated with each other. These components are materials, methods, media, and evaluation [1]. Accordingly, teaching media has become a core of learning activities. It helps teachers in providing maximum, effective, and efficient teaching [2]. Learning media that is suitable with the learning objectives and students' characteristics can improve students' learning experiences and outcomes . An interesting and fun media that can be integrated in classroom activities can be carried out by adopting some simple games [3]. Previous research has found that the use of games in the learning process can improve engagement, motivation and learning performance [4–6].

The Monopoly game is one of the simple games that can be used as learning media in the teaching and learning process [7]. The use of monopoly game in physics learning has not been widely implemented despite its potential in increasing students' involvement in learning. As stated by Firdaus [8], due to its uniqueness and interesting ways of playing, the integration of the monopoly game in science lessons could improve students' enthusiasm and confidence in learning. In line with the previous study, Irawan [9] also found that the monopoly game could foster students' enthusiasm for learning, thus the learning processes become more active, innovative, creative, effective, and fun. Since science subjects are known to be difficult and full of complex concepts, including games in learning activities could turn the lesson to be more exciting and fun. However, the monopoly games that contain science lessons, especially Physics, are still limited. Accordingly, game-based learning media in Physics learning is required.

The use of information technology (IT) in the education sector in Indonesia has become mainstream, especially these couple years due to the Covid-19 situation. Teachers and students were forced to conduct learning activities online regardless of their capacities and facilities on online learning. The utilization of Android or PC-based learning is implemented on a daily basis. Therefore, the existence of learning media based on games and software is necessary to help teachers in providing and delivering the lesson and students in acquiring and comprehending the topics being taught.

Accordingly, this study aimed to develop NaNo Physics learning media (Physics Monopoly Game) on Work and Energy topics based on software. The game adopted the rules of monopoly game in general and a study of Irawan [9]. There some modifications have been done to suit the purposes of the study. For example, if the previously developed monopoly game contains topics about cell sub-materials for high school students, we modified the contents into work and energy materials. This product could contribute to filling a scarce of game-based learning media in physics. Also, since this product was software-based, it can be utilized in any learning situation, not limited to school-learning context. Students could play the game by themselves or with their peers anytime.

2. Methods

This research was research and development (R&D) and referred to the analyze, design, development, implement, and evaluation (ADDIE) model. The procedures were a modification of the Borg & Gall development model. The research was conducted in one of the public senior high schools in Central Sulawesi. The research subjects in this study included 3 students in individual trials, 7 students in small group trials, and 14 students in field trials. Meanwhile, the validator consists of two content experts and a media expert. The object of the experiment was the quality of the NaNo Physics learning media (Physics Monopoly Game) including the learning aspects, media manufacturing aspects, and visual communication aspects.

For the sake of product assessment, the questionnaires about product quality and feasibility were distributed to the media experts, material expert, teacher, and students. The instrument used a Likert scale with 5 alternative answers, which were very good, good, enough, less, and very poor. Data were analysed quantitatively by creating a score for each response, starting from 5 points for the "very good" answer, to 1 point for the "very poor" answer. The product trial consisted of validation, one-to-one tryout, small group tryout and field-testing. The data were analyzed using descriptive statistics.

3. Results and Discussion

3.1 Learning Media Product Development

The validation of content experts obtained an average score of 5.0, with the "very good" category. It was a perfect score, thus there was no further revision made on the contents of the product. Moreover, based on the results of the media expert, the feasibility of both the media engineering and visual communication aspects were "very good" with the average score of 4.36 and 4.22, respectively. Data from individual (one-on-one trial), small-group and field-testing can be seen in Table 1.

Table 1. The trial results

Assessed Aspects	Product trials					
	individual (one-on-one)		Small-group		Field-testing	
	Score	Category	Score	Category	Score	Category
Media engineering	4.72	Very good	4.76	Very good	4.64	Very good
Visual communication	4.61	Very good	4.67	Very good	4.75	Very good
Learning	4.72	Very good	4.41	Very good	4.80	Very good

During the field trial stage, the comments given by students were generally positive. However, we decided to not follow some students' suggestions, such as adding pictures to the media. We believed that the media have had enough pictures, thus adding it would be redundant and overwhelming and could result in overlapping the provided texts. This decision was also made after considering the results of the content expert which obtained the perfect score. The detailed comparison assessments of all trials in the media engineering aspect can be seen in Figure 1. Meanwhile, the comparison of the results of the assessment of visual communication are presented in Figure 2. Overall, all indicators of media engineering and visual communication aspects were in a very good category.

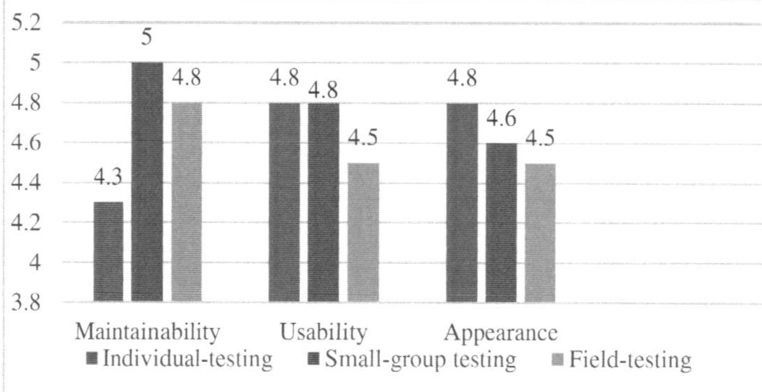

Figure 1. Average score of media engineering aspect in each indicator

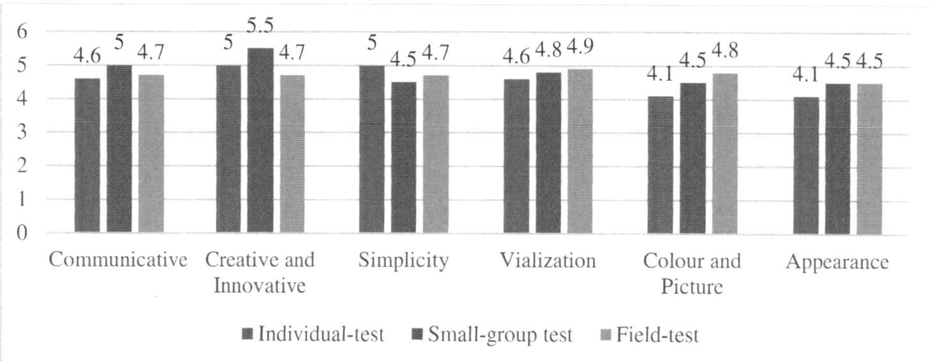

Figure 2. Average score of visual communication aspect for each indicator

The comparison of the results of the assessment of learning aspects for each indicator is presented in Figure 3. The average score of each indicator was above 4.0, which indicated that all indicators were "very good" category.

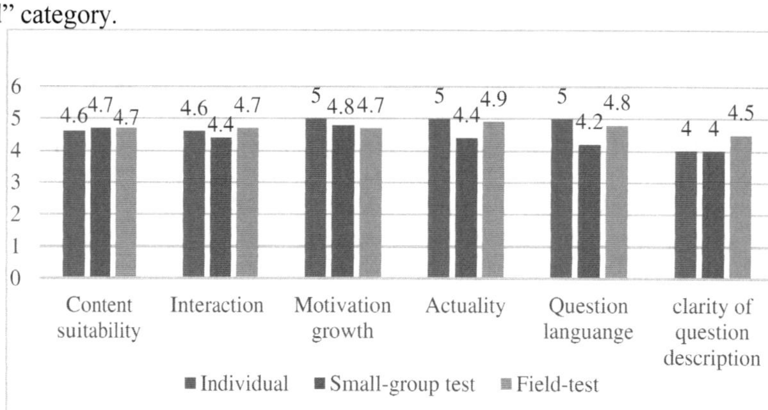

Figure 3. Average score of learning aspects f each indicator

3.2 Feasibility of the Product as Learning Media

The feasibility of the product as learning media were evaluated based on the assessment of content and media experts. The results of the media engineering aspect show an average score of 30.5 with the "very good" category. Meanwhile, the aspects of visual communication and learning resulted the average score of 33.75 and 55 in the "very good" category. These data and the comments of the experts indicated that the product has met the eligibility requirements of learning media. The appearance of final product can be seen in Figure 4.

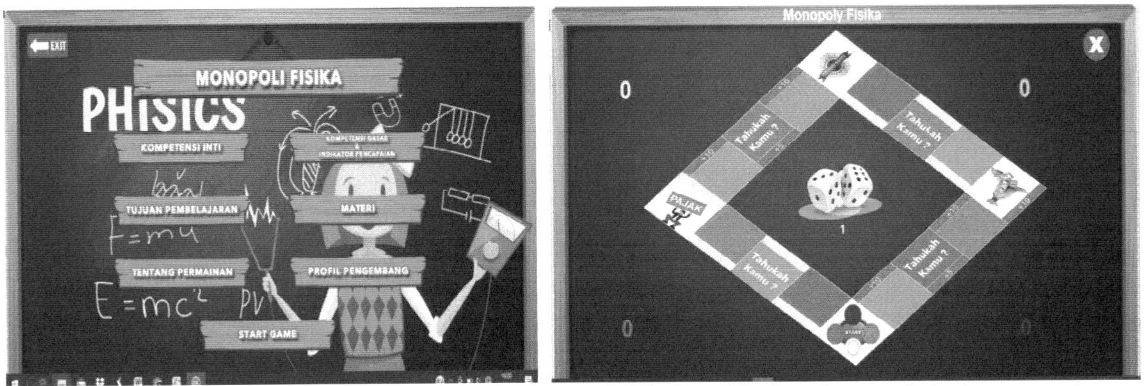

a. Cover b. Display of the board game of physics monopoly

Figure 4. Final product

3.3 Influence on students' learning motivation

Field-test was used to examine the impact of the developed product on students' learning motivation. Learning media could increase learning motivation only if the average score of each indicator has increased after its implementation. The average score of students' learning motivation increased from 3.19 in "enough" category to 4.39 in "very high". It denotes that the use of the monopoly game in physics learning can increase students' learning motivation. It is in accordance with the results of previous studies which state that game-based learning can increase students' motivation and learning outcomes [10–12]. The use of monopoly games or games in general in learning has a positive effect on learning because it allows student-centered learning [13]. In addition, the feeling that students are the center of the actor as a player unconsciously increases students' engagement with the topic being played [14,15].

4. Conclusion

Based on the results, the product of NaNo Physics learning media (Physics Monopoly Game) on work and energy concept was in the "very good" category. It means that the media is suitable for learning. In addition, its application gives a positive impact on students' learning motivation. Accordingly, the NaNo Physics learning media (Physics Monopoly Game) is recommended to be integrated in physics learning, especially in delivering the work and energy topic.

Acknowledgement

We highly acknowledge the support of the Program Kompetisi Kampus Merdeka (PK-KM) Grant of the Education, Culture, Research, and Technology of the Ministry of the Republic of Indonesia, with funding identity: 119/E1/KM.05.03/2021.

References

[1] Asnawir and Basyirudin 2002 *Media Pembelajaran* (Jakarta: Ciputat Pers)
[2] Mengajar P B 2011 *Hamalik, O* (Jakarta, Indonesia: Bumi Aksara)
[3] De-Marcos L, Domingues A, Saenz-de-Navarrette J and Pages C 2014 An empirical study comparing gamification and social networking on e-learning *Comput. Educ.* **75** 82–91
[4] Vianna Y, Vianna M, Medina B and Tanaka S 2014 *Gamification, Inc. Recreating companies through games* (Rio de Janeiro: MJV Tecnologia Ltda)
[5] De-Marcos L, Garcia-Lopez E and Garcia-Cabot A 2016 On the effectiveness of game-like and social approaches in learning: Comparing educational gaming, gamification & social networking *Comput. Educ.*
[6] Glover I 2013 Play as you learn : gamification as a technique for motivating learners *Proc. World Conf. Educ. Multimedia, Hypermedia Telemcommunications*

[7] Sukmadinata N S 2006 *Metode Penelitian Pendidikan* (Bandung: PT Remaja Rosdakarya)
[8] Firdaus Z, Zubaidah S and Sunarmi 2014 *Pengembangan Media Pembelajaran Monopoli Ipa Materi Sistem Pencernaan Makanan Untuk Siswa Kelas VIII di SMP Negeri 4 Malang* (Universitas Negeri Malang)
[9] Irawan D 2017 Pengembangan Media Pembelajaran Monopoli Fisika pada Materi Besaran dan Satuan *J. Alumni Pedidikan Fis.* **3** 247–54
[10] Trevino R, Majcher C, Rabin J, Kent T, Maki Y and Wingert T 2016 The effectiveness of an educational game for teaching optometry students basic and applied science *PLoS One*
[11] Satrio Y D, Wardoyo C, Sahid S, Fauzan S and Ma'ruf D 2021 The Effectiveness of Educational Games on Post-Pandemic Learning *KnE Soc. Sci.*
[12] Zeng H, Zhou S N, Hong G R, Li Q Y and Xu S Q 2020 Evaluation of interactive game-based learning in physics domain *J. Balt. Sci. Educ.*
[13] Huang Y M, Huang S H and Wu T T 2014 Embedding diagnostic mechanisms in a digital game for learning mathematics *Educ. Technol. Res. Dev.*
[14] Faiola A, Newlon C, Pfaff M and Smyslova O 2013 Embedding diagnostic mechanisms in a digital game for learning mathematics Embedding diagnostic mechanisms in a digital game for learning mathematics Embedding diagnostic mechanisms in a digital game for learning mathematics Computers *Hum. Behav.* **29** 1113–21
[15] Schrader C and Bastiaens T J 2012 The influence of virtual presence: Effects on experienced cognitive load and learning outcomes in educational computer games *Comput. Human Behav.*

The Development of Local Disasters-based Mitigation Module Integrated to Physics Learning

Yuliani* **and Wahyono**

Sub-Department of Physics Education, Faculty of Teacher Training and Education, Tadulako University, Jl. Soekarno Hatta KM. 9 Kampus Bumi Tadulako Tondo Palu-Sulawesi Tengah, Indonesia

* yulianiy13@gmail.com

Abstract. This study aimed to develop a local disaster potential-based mitigation module that is integrated into physics learning. The compilation of these teaching materials can be utilized as one of the learning media for students to learn about disaster mitigation materials independently. This study was research and development using 4D model that consisted of 4 phases namely, Define, Design, Development, and Disseminate. The participants in this study were 12 students from SMPN 1 Palu. The feasibility test phase of the product was conducted to determine the feasibility of the instructional materials. The trial phase was conducted on trials of material experts, media experts, teachers, and students. Based on the results of trials and questionnaire analysis, the average score of the material experts assessment was 3.24 categorized as "Good", the average score of media expert assessment was 3.00 categorized as "Good", the average score of teacher assessment was 4 0.0 categorized as "Very Good", and the average score of the assessment based on student responses was 3.52 categorized as "Very Good". The result of the study shows that the teaching materials generated, particularly those on disaster mitigation, are very suited for use in the learning process.

1. Introduction

Physics is a way of looking at the universe, comprehending how it works and how its various components interact. Students cannot be separated from studying the symptoms or phenomena that occur in the universe, especially those that are near to everyday life, when studying physics. In conclusion, learning physics is a process of gaining information through the study of diverse physical processes that occur throughout the universe [1].

Natural phenomena are non-artificial events in the eyes of physics, which means that they are not caused by humans, despite the fact that they can have an impact on humans. Natural disasters are intimately linked to environmental factors. Natural disasters caused by human activity can be avoided by protecting the environment. Recognizing and maintaining the area's local potential is one strategy to conserve the environment [2].

Palu City is on of cities in Indonesia that is prone to natural disasters. Earthquake, Tsunami, and Liquefaction are the natural disasters that have struck Palu. Delivering disaster mitigation items is an approach to combat these natural disasters. One of the fields that is applied to avert disasters is science [3,4].

Learning about disaster mitigation issues can be delivered through a variety of teaching materials, one of which is printed teaching materials in the form of modules. A module is a learning package that is linked to a certain unit of study content [5]. The modules are organized in a logical and appealing manner, and they provide content, methodologies, and evaluation that can be utilized independently [6]. According to the research findings of Pambudi [7] modified modules incorporating natural disaster materials into science lessons are appropriate for use and can improve students' learning outcomes.

According to an interview with one of the science subject teachers at SMP Negeri 1 Palu, the students at the school had never been taught about science learning related to disaster mitigation and had never used a module on science learning related to disaster mitigation; instead, they only use textbooks that teach about science learning in general, with no disaster mitigation material. To address this issue, modules should be used as teaching materials to stimulate students' interest in learning environmental science in order to lessen the danger of natural disasters. Modules are contextually structured and can attract students.

2. Methods

This research was Research and Development (R&D) is a research model that is used to generate products and to evaluate their efficacy [8]. It utilized 4D model that consists of 4 phases namely Define, Design, Development, and Disseminate.

In define phase, the researcher established and defined the learning criteria at this point. In addition, the researcher conducted observations in schools that covered five primary steps: initial and final analysis, student analysis, concepts analysis, task analysis, and learning objective analysis. Meanwhile, in design the researchers designed the product. It covered four steps, namely the tests preparation, media selection, format selection and initial design. The next phase is the development stage. This phase aimed to produce product development which is carried out through two steps, namely expert assessment (materials and media) followed by revision and development trials (students). The last phase was the dissemination. This stage aimed to promote the development product so that it can be accepted by users, whether individuals, groups, or systems. At this phase, the module developed was distributed to several agencies in Palu city including SMP Negeri 1 Palu, the the National Disaster Management Coordinating Board (BPBD) office in Palu City, supervisor and reviewer, and the library of the Physics education study program to determine the quality of the disaster mitigation module. For limited test, 12 students of 8th grade were involved as research subjects to assess the feasibility of teaching materials.

The instrument utilized in this study was a questionnaire which aimed to assess the feasibility of the learning content/material and media. The questionnaire was distributed to material experts, media experts, teachers to assess the feasibility of the module and questionnaire that was distributed to students as research respondents. The questionnaire was a 4-scale of Likert scale. For the data analysis, descriptive statistic was used to calculate the average value of each aspect. This average value analysis technique was determined based on Arikunto's opinion [9] which states that to find out the final score rating for each item of the research questionnaire, the total score obtained is divided by the number of respondents who answered the assessment questionnaire. The average value was by numbering the options from the strongly disagree to strongly agree option sequentially from 1 to 4. The average scores from the evaluation of expert validators, teachers and students then were converted into qualitative form based on Table 1 [10].

Table 1. Product value criteria

Average Score	Criteria
3,25 $<\bar{X}\leqslant$ 4,00	Very Good (SB)
2,50 $<\bar{X}\leqslant$ 3,25	Good (B)
1,75 $<\bar{X}\leqslant$2,50	Less (K)
1,00 $\leqslant\bar{X}\leqslant$ 1,75	Very Less (SK)

The assessment score or the feasibility level for both each aspect and the whole of the module as a learning medium utilized Tables 1 and 2 above as a reference for assessing data generated from the material experts and media experts validations, feasibility tests for use by teachers and trials on students to make it easier to determine a value category whether or not the generated module is feasible to be used as a learning media.

3. Results and Discussions

3.1. Content Validation

The material expert is one of the physics lecturers at Tadulako University who assesses the material presentation, linguistic, and educational values feasibilities. The detailed scores for each aspect are provided in Table 2, 3 and 4. Overall, the average score of content assessment was 3.24 and is in the "Good" category.

Table 2. Validation results of materials presentation feasibility aspects

Assessment Aspect	Score	Category
The material presented is suitable/relevant to the applicable learning.	4.00	Very Good
The concepts and definitions presented are in accordance with the existing concepts and definitions in physics.	4.00	Very Good
The description of the material presented is logical (reasonable), coherent, and beneficial for SMP students.	3.00	Good
The materials and examples presented support independent learning for SMP students.	3.00	Good
The examples presented are in accordance with the existing conditions in the surrounding environment.	3.00	Good
Average	3.40	Very Good

Table 3. Validation results of linguistic presentation feasibility aspects

Assessment Aspect	Score	Category
The language used is in accordance with the students' thinking level.	4.00	Very Good
The language used encourages students' curiosity to finish learning the material.	3.00	Good
The language use remains polite and does not reduce educational values.	3.00	Good
Average	3.33	Very Good

Table 4. Validation results of educational value feasibility aspects

Assessment Aspect	Score	Category
The media is easy to use in the student learning process both independently and in class.	300	Good
The media is able to increase students' motivation in learning about disasters and how to overcome these disasters.	3.00	Good
The media is able to expand students' insight in dealing with disasters.	3.00	Good
The media can assist teachers to provide references about the relationship between physics and science and technology development.	3.00	Good
Average	3.00	Good

3.2. Media Validation

The media expert is one of the employees of the Palu City Regional Disaster Management Agency (BPBD) who assesses presentation quality feasibility, design feasibility, and implementation feasibility. The average score for all aspects was 3.00. it indicates that the feasibility assessment by media expert was in the good category. The data can be seen in tables 5, 6, and 7.

Table 5. Validation results of display quality feasibility aspects

Assessment Aspect	Score	Category
The language used is in accordance with the students' thinking level and is easy to understand.	3.00	Good
The font size and color used are proportional and clear.	3.00	Good
The presentation of images and disaster information is attractive and proportional.	3.00	Good
Average	3.00	Good

Table 6. Validation results of design feasibility aspects

Assessment Aspect	Score	Category
Media display is attractive.	3.00	Good
The media can be an option to obtain knowledge about the process of disaster (Earthquake, Tsunami, and Liquefaction) and the signs of a disaster and reduce the burden on teachers.	3.00	Good
The media design has its own attractiveness that makes students want to read and use it in the learning process.	3.00	Good
Average	3.00	Good

Table 7. Validation results of implementation feasibility aspects

Assessment Aspect	Score	Category
The disaster mitigation media can encourage students' curiosity on how to take action when a disaster occurs.	3.00	Good
The disaster mitigation module is easy to use in learning both inside and outside the classroom.	3.00	Good
The disaster mitigation module supports students to learn independently.	3.00	Good
The disaster mitigation module increases students' enthusiasm in learning about disasters, especially about the occurrence of disasters (Earthquake, Tsunami, and Liquefaction) as well as efforts to reduce the risks of these disasters.	3.00	Good
Average	3.00	Good

3.3. Limited-test

The limited test was carried out on 12 students of SMPN 1 Palu, the students were provided a questionnaire containing 13 statements and there were 4 categories of choices from strongly agree to strongly disagree. The result obtained from the overall average value of this limited test is 3.52 and is in "very good" category. Since the score is in very good category, the module is suitable for use as a student learning media. The data can be seen in Table 8.

Furthermore, the assessment results of the overall feasibility test conducted by one of the physics teachers from the same school obtained a perfect score for all aspects. The aspects included the feasibilities of content, material, design, module display and module completion. Therefore, the average result of the feasibility test of overall assessment was 4.00 and is in the "Very Good" category. It shows that that this module of disaster risk reduction is suitable for further use in learning in SMP without any revision. The data can be seen in Table 9.

Table 8. Analysis result of average rate based on students' responses

Statements	Score
I become more excited to learn physics with the disaster mitigation module than just listening to the explanations about natural disasters in general.	3.25
I like the display of this module media design because it has a harmonious and attractive composition, color, font, image, and design.	3.41

In my opinion, using this disaster mitigation module can make physics lessons less boring.	3.16
This disaster mitigation module can be an alternative learning resource.	3.75
Guidance such as a disaster mitigation module is required so that the community is more prepared for disasters.	3.75
The disaster mitigation module presented encourages you want to learn more about the process of Earthquake, Tsunami, and Liquefaction.	3.58
The module media presented are interesting both in material and in design.	3.25
The disaster mitigation module can expand your knowledge about the process of earthquake, tsunami and liquefaction disasters. Therefore, you become more prepared to face these three disasters.	3.91
Every material in the disaster mitigation module is understandable and interesting to read.	3.50
The disaster mitigation module media requires further development.	3.50
The image design of the disaster mitigation module is a special attraction for readers.	3.25
The module media presented can evoke your interest to participate the efforts to reduce the risk of earthquakes, tsunami and liquefaction.	3.75
The disaster mitigation module media can provide information on how to take action when an Earthquake, Tsunami, and Liquefaction disaster occur.	3.75
Average	3.52

Table 9. Analysis result of teacher feasibility average

Indicators	Score	Category
The attractiveness of the module cover.	4.00	Very Good
The orderliness of cover page design.	4.00	Very Good
The selection of type and font size makes the media more attractive.	4.00	Very Good
The transitions continuity between pages	4.00	Very Good
The easiness to read the written text.	4.00	Very Good
Color selection	4.00	Very Good
The conformity of stories, images and materials	4.00	Very Good
The completion of module.	4.00	Very Good
Average	4.00	Very Good

4. Conclusions

Overall, based on the evaluation of expert validators and prospective users, the module for disaster risk reduction (DRR) with the context of Palu, Central Sulawesi in learning physics is feasible to be used as one of the teaching media.

Acknowledgement

We highly acknowledge the support of the Program Kompetisi Kampus Merdeka (PK-KM) Grant of the Education, Culture, Research, and Technology of the Ministry of the Republic of Indonesia, with funding identity : 119/E1/KM.05.03/2021.

References

[1] Young H D and Freedman R A 2003 *Fisika universitas edisi kesepuluh jilid 1* (Jakarta: Erlangga)
[2] Ibrohim I 2015 *Prosiding Semnas dan Enterpreneurship* **2** 1–19
[3] Hening W R N, Sudarmin S and Mustikaningtyas D 2013 *Unnes Sci. Educ. J.* **2** 254–261

[4] Masfuah S, Rusilowati A and Sarwi S 2011 *J. Pendidik. Fis. Indones.* **7** 115–120
[5] Ayriza Y 2009 *J. Kependidikan Penelit. Inov. Pembelajaran* **39** 141–156
[6] Setyowati R, Parmin P and Widiyatmoko A 2013 *Unnes Sci. Educ. J.* **2** 245–53
[7] Pambudi D I 2015 *Elem. Sch. J. Pendidik. dan Pembelajaran ke-SD-an* **2** 22-36
[8] Sugiyono S 2010 *Statistika untuk penelitian* (Bandung: Alfabeta)
[9] Arikunto S 2006 *Metode Penelitian Kualitatif* (Jakarta: Bumi Aksara)
[10] Widoyoko E P 2014 *Teknik penyusunan instrumen penelitian* (Yogyakarta: Pustaka Pelajar)

Developing android-based teaching material on temperature and heat using ADDIE model

T Sriwahyuni[1]*, Kamaluddin[1], and S Saehana[1]

[1]Sub-Department of Physics Education, Faculty of Teacher Training and Education, Universitas Tadulako, Jl. Soekarno Hatta KM. 9 Kampus Bumi Tadulako Tondo Palu-Central Sulawesi, Indonesia

*Corresponding author: ririnfis17@gmail.com

Abstract. Physics teaching materials are essential in supporting students to solve the problem and understand abstract physics concepts, if they are attractively designed. In this study, the researchers aim to create android-based teaching material on temperature and heat to increase students' interest and motivation. This study is the development research using the ADDIE model, which involves 5 stages. They are Analysis, Design, Development, Implementation and Evaluation. The research subjects were 20 students of class XI. The test and questionnaire analysis results indicated that the average score of the two validators was 3.66 and was categorized as "Very good." Assessment by the teacher depicted an average score of 3.45 categorized as "Very Good," and the results of the analysis of student response assessment obtained an average score of 3.25 which means "Strongly Agree". The results showed that the android-based teaching material developed is feasible as teaching material for students.

1. Introduction

The rapid development of science and technology has impacted almost every aspect of lifestyle today, and education is no exception. The 2013 curriculum has required teachers to possess the competence to master and utilize ICT (information and communication technology) to facilitate students in the learning process to be more effective and efficient [1]. The expected output of ICT-based learning is that students can compete and face global challenges in the future [2].

According to Shamim and Raihan, the use of ICT in education can be divided into two categories: ICT for education and ICT in education [3]. ICT for education refers to the development of ICT for learning, while ICT in education is the adoption of several components of ICT for use in the learning process. Referring to the statement, the teacher is free to choose what strategies will be used to integrate ICT in the classroom. However, it should be realized that although ICT is increasingly popular, not a few schools are still facing challenges, so they have not implemented ICT as a whole [4,5].

This is as found in SMAN 1 Tombusabora where teachers have not shifted their role as the primary source of information. For example, in the physics learning process, the teaching materials are still dominated by printed books. Those kinds of teaching materials will decrease students' learning motivation because they tend to deal with formulas that are not attractively presented. Based on these conditions, researchers need to take advantage of the use of ICT to develop digital teaching material in the form of applications. Thus, students are motivated to learn, and teachers can be assisted by supportive learning resources to improve the quality of physics learning.

The technology that will be used as a media of digital teaching material is an Android-based smartphone. In fact, most students of SMAN 1 Tombusabora already have a smartphone. However, its use is more often for games and social media. Thus, researchers think it will be more efficient if the application can be installed on students' smartphones. This study is in line with a study conducted by Maulida et al which states that android-based teaching materials are feasible to use in learning and potential to improve students' critical thinking skills [6].
The topics that will be integrated in this teaching material are temperature and heat. The choice of temperature and heat material is also based on the research of Setyadi and Komalasari, which showed that 36.3% of students' mastery of temperature and heat material is still in the poor category [7]. Furthermore, according to research conducted by Febrianti et al., high school students had misconceptions about temperature and heat material by 57.85%. [8] Therefore, the purpose of developing digital teaching material is to produce an Android-based application that is easy to use and has an attractive appearance. This visual teaching material helps students introduce, enrich knowledge, and clarify the concept of temperature and heat material. In addition, this product is easy to operate for both students and teachers.

2. Method

2.1 Research Setting

This research was conducted in SMAN 1 Sindue Tombusabora. This school is one of the schools located in Donggala Regency, Central Sulawesi Province. The time needed for the development of teaching material to implement the product feasibility test is 9 months. The respondents in this study involved 20 students of class XI SMA Negeri 1 Sindue Tombusabora

2.2 Research Design

The design and development of the application in this study are based on the ADDIE model. The choice of this model is because the ADDIE model has systematic and easy-to-follow stages of creating a product [9]. According to Aldoobie (2015), this model is an approach that helps instructional designers, content developers of any kind, or even teachers to create efficient and effective teaching designs [10]. The ADDIE model consists of several phases: Analysis, Design, Development and Implementation and Evaluation. Concrete steps to develop android-based teaching material are described as follows:

1. At the analysis stage, both literature and field studies were conducted. Field study in the form of observation was carried out by identifying the students' and teachers' needs in the physics learning process. A literature study was carried out by analyzing the curriculum and finding references to solve the problems found.
2. Design stage is the stage to design a strategy for solving problems in the physics learning process. At this stage, the researchers designed an initial overview of Android-based teaching material by formulating the flow of material delivery in the form of flowchart, making storyboards, selecting applications, and preparing the tools and material used.
3. Development stage is the stage of creating the design of teaching material into an application. Once the product finished, it was validated by two expert lecturers who mastered the content of the topic chosen, to assess the feasibility of the product.
4. Implementation stage is the stage where the product that had been developed was tested on the assigned respondents and teachers. Then they were asked to provide responses using research instruments.
5. Evaluation stage is the assessment stage, which refers to the feasibility test results from validators, teachers, and students. At this stage, the feedback on the product was used as a guide for improving the quality of the product.

2.3 Data Collection Techniques

The study uses questionnaires to obtain the data. Questionnaire is one of the instruments containing a list of questions or statements to find out the respondents' opinions regarding a product or social phenomenon [11]. The type of questionnaire in this study is a closed questionnaire because respondents were only given the option to fill in a checkmark in the alternative answer column. However, to give feedback on the product, the researchers provided a comment column at the end of the questionnaire. The questionnaire assessment uses a Likert scale measurement with four alternative answers ranging from the most positive to the most negative. In need of quantitative data processing, each answer choice from the statement is given a score. The obtained data is analyzed by calculating the average value of the overall score [12]. Data from the questionnaire was then converted qualitatively, as in Table 1 [13].

Table 1. Likert scale category for experts and teachers

Average Score	Criteria
$3.25 < \underline{X} \leqslant 4.00$	Very Good (VG)
$2.50 < \underline{X} \leqslant 3.25$	Good (G)
$1.75 < \underline{X} \leqslant 2.50$	Poor (P)
$1.00 \leqslant \underline{X} \leqslant 1.75$	Very Poor (VP)

The technique of analyzing student data is similar to that of experts and teachers, but what is different is the qualitative interpretation. Experts and teachers were asked to evaluate the quality of the product, by presenting the material and the overall appearance of teaching material. The students were asked for approval to evaluate whether the product developed could improve their understanding of the physics material or vice versa. Therefore, the mean score obtained from the quantitative analysis can be converted qualitatively by referring to Table 2.

Table 2. Likert scale category for students

Average Score	Criteria
$3.25 < \underline{X} \leqslant 4.00$	Strongly Agree (SA)
$2.50 < \underline{X} \leqslant 3.25$	Agree (A)
$1.75 < \underline{X} \leqslant 2.50$	Disagree (D)
$1.00 \leqslant \underline{X} \leqslant 1.75$	Strongly Disagree (SD)

3. Results and Discussion

The analysis stages in the ADDIE model revealed that the teaching material used at SMAN 1 Sindue Tombusabora had not integrated ICT in the physics learning process. In fact, integrating ICT in learning can also improve student literacy in technology. The delivery of physics material, especially the material on temperature and heat, had been dominated by printed books in the form of textbooks or worksheets. The teaching method also still employed a teacher-centered approach. This means that the teacher tends to be the primary source of knowledge in the classroom rather than a facilitator who supports student learning activities. Thus, it reduces students' interest in learning physics. Moreover, for students, the physics subject is considered difficult, unattractive, unpleasant, and even scary for some students [14]. Therefore, researchers need to create Android-based teaching material on temperature and heat. In general, all students in SMAN 1 Sindue Tombusabora already have smartphones. This fact has become a reference for researchers to create android-based teaching material that can be accessed via smartphones to increase student motivation to learn. The material that will be integrated into the teaching material is temperature and heat. According to the 2013 curriculum, the material was discussed in class XI.

Based on the analysis results above, android-based teaching material was developed at the design and development stage. At the design stage, the outline of the material, instructional objectives according to the syllabus and the software used had been set. The researchers also designed the appearance of the application, including the selection of visual colors and figures. Then at the development stage, the product was developed. Android-based teaching material was composed using Game Maker Studio v1.4 software. Even though the name is Game Maker, it can be used to create applications other than games and the manufacturing process without complicated scripts. The researchers collected all materials from literature review, both from books and other references, then inserted them into the software. The results of android-based teaching material can be seen in Figure 1.

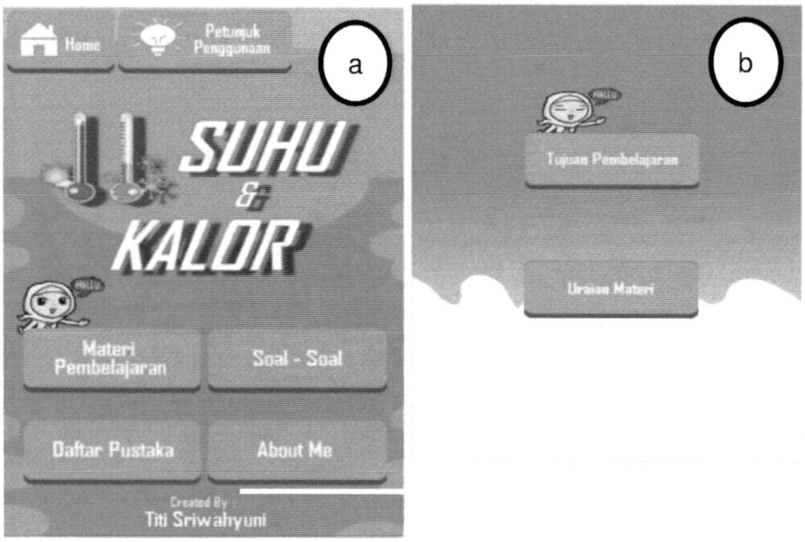

Figure 1. Display menu for Android-based teaching material

Figure 1.a is the initial display when the user opens the application for the first time. Students will be presented with a home menu with features such as instructions for use, learning material, questions, bibliography, and about me. Students can pick the user manual icon to learn the instructions for using the application. Furthermore, the temperature and heat material can be accessed by pressing the learning material button and will enter the display as shown in Figure 1.b. The menu of questions serves to evaluate students' understanding of the material accompanied by discussion. The table of contents menu contains a collection of references from the subjects in the teaching material, while the about me feature describes the application makers. Before being implemented to respondents, the product requires a feasibility test by experts. Questionnaires were given to two lecturers whose assessment results are shown in Table 3.

Table 3. The results of the product feasibility assessment by experts

No	Assessment Aspects	Score	Category
1	Material Feasibility Aspect	3.25	Very Good
2	Feasibility of the Effect of Teaching Material on Learning Strategies	3.40	Very Good
3	Presentation Feasibility	4.00	Very Good
4	Media	4.00	Very Good
	Total	3.66	Very Good

As Table 3 shows, the experts gave a total score of 3.66, categorized as "very good." In fact, there were two aspects of the assessment that had a perfect score, the media aspect, and presentation feasibility. When referring to the statements in the questionnaire, the assessment indicators for these two aspects were about the flow of material that could be easily understood and the design to motivate students to learn using Android-based teaching material. Yet, there were some additional notes for researchers for improving the teaching material. The respondents suggested that researchers should add the explanation of the expansion of liquids and gases and add the example of heat transfer in everyday life. Based on these suggestions, the researchers made improvements according to the suggestions given. The improved Android-based teaching material was then tested on teachers and students. Tests on teachers and students were included in the implementation phase of the ADDIE model. Similar to the assessment carried out by experts, teachers were also asked to provide an assessment. However, the list of statements in the questionnaire was different. The results of the teacher's product feasibility test assessment are depicted in Table 4.

Table 4. The results of the product feasibility assessment by the teachers

No	Indicator	Score	Classification
1	Curriculum	3.00	Good
2	Contents	3.50	Very Good
3	Product Feasibility Evaluation	4.00	Very Good
4	Languages	3.33	Very Good
	Average	3.45	Very Good

The average score from the above assessment results shows a score of 3.45. If converted according to the Likert scale measurement, this product is considered very good and suitable for the physics learning process, especially on temperature and heat subject. This is also supported by the absence of additional suggestions or criticisms in the questionnaire comments column. Therefore, the feasibility test continued giving questionnaires to 20 students as respondents. The questionnaire for students has 13 positive statement items. Each statement item from each respondent was analyzed to get an average score. The final score is 3.50, which is categorized as strongly agree. More clear explanation can be seen in Table 5.

Table 5. The assessment result of student responses to the product

No.	Statement	Score
1	I get excited and motivated to learn physics with the presence of android-based teaching material	3.90
2	I can study independently with teaching material	3.30
3	I can learn according to my speed and intensity of learning	3.15
4	I prefer to learn with android-based teaching material rather than listening to the teacher's explanation in class	2.95
5	Through android-based teaching material I can understand the material	3.30
6	I think this teaching material can help the teacher to explain the material	3.65
7	In my opinion, using this application makes physics subjects not boring	3.65
8	Sample questions and pictures in the application of teaching material help me understand well about temperature and heat	3.65
9	I like the design of android-based teaching material because it has a match and attractive color and image composition	3.50
10	Instructions for using android-based teaching material make it easier for me to use the application	3.40
11	Android-based teaching material motivates me to learn physics	3.65
12	I think this teaching material can make physics learning flexible because it can be studied anywhere and anytime	3.70
13	I think the writing in the application is clear and easy to read	3.75
	Average Total Score	**3.50**

Table 5 shows the positive responses of students when using teaching material. It can be seen from the score for the statement item about students' motivation to give the highest score of 3.9 which means that this android-based teaching material can motivate students to learn physics. In addition, several statement items about the role of teaching material in improving students' understanding of temperature and heat material also scored in the category of strongly agree. However, on the other hand, there was an average score of 2.95, categorized as agree. This shows that students also still need the teacher's role in the physics learning process because the teaching material in the learning process only act as a supporting learning resource for teachers in the classroom. Then the final stage of the product being developed is the evaluation stage. According to Murdiono et al. [15], this stage was a phase to see if the success of teaching material can be useful for students or vice versa. The measure of the product's usefulness can be seen from the overall analysis of the questionnaire given to experts, teachers, and students. Overall, it can be said that these teaching materials are suitable for use in learning physics.

Several studies on the development of android-based teaching material had also been carried out by other researchers, but there are differences including: (1) the topics discussed in the teaching material were different. (2) Applications that have been developed require more than one application such as Adobe flash CS6, CorelDrawX7, Adoobe Air 3.5 and macromedia flash. (3) Practice questions in limited applications [16,17].

This development of Android-based teaching material has several advantages: (1) Teaching material in this Android application can be used by students on any smartphone without an internet connection or accessed offline. (2) There is a review of the discussion of practice questions. Reviews of discussion questions are useful to help students understand the flow of answers to practice questions. (3) The capacity of this application is relatively low, which is only 27.64 MB. Thus, the app installation will not occupy the storage of the device. However, the drawback is that the physics formulas in the application must be converted to image format (JPG) first.

4. Conclusion

The design and development of the application in this study have produced an android-based teaching material product. The development steps are based on the ADDIE model, consisting of 5 stages: analysis, design, development, implementation, and evaluation. Based on the feasibility test of the expert teachers, the students gave a score of 3.66, 3.45, and 3.50, respectively. The assessment results showed that this teaching material can be used as one of the teaching material to increase students' interest and motivation to learn physics. Because the dissemination of this product is still on a small scale, in the future, this product can be tested for effectiveness in a broader scale.

Acknowledgement

We highly acknowledge the support of the Program Kompetisi Kampus Merdeka (PK-KM) Grant of the Education, Culture, Research, and Technology of the Ministry of the Republic of Indonesia, with funding identity: 119/E1/KM.05.03/2021.

References

[1] Kemendikbud 2014 *Implementasi Kurikulum 2013* (Jakarta: Press Workshop)
[2] Siregar Z and Marpaung T B 2020 *Bio. Educ. Sci. Technol. J.* **3** 61 – 69
[3] Shahim R H and Raihan A 2016 *Int. J. Vocat. Tech. Educ.* **8** 12–19
[4] Nikolopoulou K and Gialamas V 2016 *J. Comput. Educ.* **3** 59–75
[5] Rivalina R and Siahaan S 2020 *J. Teknodik* **2** 73–87
[6] Maulida H, Sinaga P and Susilawati M 2019 *J. PETIK* **5** 70–76
[7] Setyadi E and Komalasari A 2012 *Berk. Fis. Indones. J. Ilm. Fis. Pembelajaran dan Apl.* **4** 46–49
[8] Febrianti J, Akhsan H and Muslim M 2019 *J. Inov. dan Pembelajaran Fis.* **6** 90–102
[9] Patel S R, Margolies P J, Covell N H, Lipscomb C and Dixon L B 2018 *Front. public Heal.* **6** 113
[10] Aldoobie N 2015 *Am. Int. J. Contemp. Res.* **5**
[11] Roopa S and Rani M S 2012 *J. Indian Orthod. Soc.* **46** 273–277
[12] Arikunto S 2013 *Prosedur Penelitian Suatu Pendekatan Praktik* (Jakarta: Rineka Cipta)
[13] Widoyoko E P 2014 *Teknik penyusunan instrumen penelitian* (Yogyakarta: Pustaka Pelajar)
[14] Halim A and Lestari D 2019 *J. Phys. Conf. Ser.* **1280** 52060
[15] Murdiono M 2020 *Int. J. Interact. Mob. Technol.* **14** 180–193
[16] Sukrisna D and et al 2018 *Seminar Nas. Sains dan Teknol.* **1** 1–10

[17] Shabrina 2018 *Int. J. Instr.* **11** 287–302

Langmuir and Freundlich isotherm equation test on the adsorption process of Cu (II) metal ions by cassava peel waste (*Manihot esculenta crantz*)

S Nuryanti[1*], Suherman[1], S Rahmawati[1*], M Amalia[1], T Santoso[1] and H Muhtar[1]

[1]Chemistry Study Program, Mathematics and Natural Science Division, Faculty Of Teacher Training And Education, Tadulako University, Jl. Soekarno-Hatta Km 9 Palu 94118 Indonesia.

[*]Corresponding author: sittirahmawati@yahoo.com

Abstract. This Study aims to determine the adsorption process of Cu (II) metal by cassava peel waste (*Manihot esculenta crantz*) meeting the Langmuir equation and Freundlich. Research method cassava peel made into activated carbon, then determine the mass, pH and optimum contact time. Then the determination of the maximum adsorption capacity was carried out by testing with the Langmuir isotherm and Freundlich isotherm equations. The results showed that the best conditions (optimal conditions) were obtained with the addition of 0.5 grams of active carbs from cassava peels. The percentage of Cu ion adsorption in these conditions was 97.72%, at pH 6 and a contact time of 60 minutes and the absorption capacity was 98.49%. The maximum adsorption capacity of cassava peel activated carbon to Cu(II) ions at the optimum condition was determined based on the Langmuir and Freundlich isotherm equations. The results obtained were -51.813 mg/g and 26,792 mg/g, respectively.

1. Introduction

Along with the times and accompanied by population growth and industrial development, it triggers an increase in the amount of industrial waste that can cause environmental pollution. One of them is water pollution by waste containing heavy metals, such as Cu. The impact of being contaminated with Cu metal in the body causes stomach cramps, nausea, vomiting, diarrhea, kidney failure, liver and even death [1]. Copper metal (Cu) is found in water and soil both in the form of ions and compounds. Naturally Cu enters the waters through erosion, rock erosion, or in the atmosphere brought down by rainwater [2]. Although heavy metal Cu (II) is an essential element for the body, it is in an amount that exceeds the threshold because with industrial developments it can be harmful to health.

Along with the development of the industry, various efforts were made to reduce pollution caused by these heavy metals. One of the methods that have been developed to remove heavy metals is adsorption [3]. Adsorption is the process of absorption by certain solids against certain substances that takes place on the surface of the solid due to the attraction of atoms or molecules on the surface of the solid without seeping into it. The solid surface is referred to as the adsorbent, whereas the adsorbed substance is referred to as the adsorbate [4]. The adsorbent which is generally used is activated carbon. Activated carbon is an amorphous compound derived from materials containing carbon or charcoal which are specially treated to obtain high absorption energy [5].

The adsorption process is currently being developed by utilizing waste containing carbon, including banana peels, coffee peels, rice husks, cocoa peels, durian peels. Besides this waste, cassava peel is also a material that contains carbon and has the potential as an adsorbent. Currently, this waste has not been utilized optimally, while the potential availability of cassava peels is very abundant. Based on survey data, cassava peel waste in the city of Palu, Central Sulawesi, reached 147 kg/per day. One of the producers of cassava peel waste which is quite large is the cassava chip manufacturing industry.

The superiority of cassava peel compared to durian peel waste, banana peel, cocoa peel and coffee peel as an adsorbent because it contains lignocellulose which has the potential to adsorb cationic waste. Lignocellulose is a natural material consisting of hemicellulose, cellulose and lignin. The main functional groups on the surface of lignocellulosic materials are hydroxyl, aldehyde, carboxylate, and groups Cyano [6]. Therefore, materials containing lignocellulose, such as cassava peel, become an alternative as an adsorbent that can be utilized, because of its large availability and valuable because it is only used as waste.

Research data from cassava peel contains carbon (C) of 59.31%, hydrogen (H) of 9.78%, oxygen (O) of 28.74%, nitrogen (N) of 0.11%, and water (H_2O) by 11.4% [7]. Based on the high carbon content, cassava peel has the potential to be used as raw material for activated carbon with an activation and carbonization process that can be used as an adsorbent [8].

Activated carbon from cassava bark has been used for testing the absorption of Fe ions and obtained the highest absorption of 51.97% at the carbonization temperature of 600°C [9]. The same study used activated carbon from corn cobs as Cu (II) ion adsorbent. The results showed that the adsorption of Cu (II) metal occurred at an optimum pH of 7 with an absorption of 98.34%, the optimum contact time occurred at 60 minutes with the percentage of Cu (II) adsorbed at 96.37%, the maximum adsorption capacity of Cu (II) ions of 2.4160 mg/gr [10]. Based on the results of the research on the adsorption of cassava peel activated carbon to Fe (II) ions and corn cobs activated carbon adsorption to Cu(II) ions, in this study, the optimum pH determination was used variations of 4-9, the mass of active carbohydrates was 0.3-0.6 gram and variation of contact time 15, 30, 60 and 90 minutes. These variations were used to absorb Cu(II) ions using activated carbon from cassava peels, while for testing the absorption capacity in this study the Langmuir and Freundlich isotherm equations were used.

2. Methods

The materials used in this study were cassava peels obtained from local traditional markets. Other chemicals such as 0.1 N NaOH solution, 0.1 N HCl solution, solid CuSO4.5H2O, were obtained from E. Merck and used without further purification.

2.1. Sample Preparation

The Cassava peel was washed thoroughly and separated from the brown outer skin then cut into small pieces and then dried in the sun for 2 days [11].

2.2. Making Adsorbent

Dry cassava peels are put into a furnace to make charcoal at 400°C for 1 hour and then allowed to cool at room temperature. The cassava peel charcoal was mashed using a blender until the charcoal passed through a 100 mesh sieve. Cassava peel drying and carbonation in this process aims to decompose hydrocarbon compounds such as cellulose and hemicellulose into pure carbon. Carbonated charcoal was ground and sieved using a 100 mesh sieve. With the purification process, the chemical activation process will be more evenly distributed because the smaller the size of the cassava peel carbonation charcoal, the greater the surface area that will be in contact with the activator so that more activated carbon and more pores are formed on each carbon. particles [12].

2.3. The Activation Process of Cassava Peel

The activation process is carried out by chemical means, namely cassava peel charcoal soaked in 0.1 N NaOH solution for 60 minutes. Then the activated carbon produced was washed with aquadest and 0.1 N HCl until the filtrate had a neutral pH (pH 6 to 7) which was measured using a pH meter. Then filtered using filter paper. Samples obtained dried in an oven at 1100C for 60 minutes [12]. Activation of cassava peel charcoal aims to increase metal absorption, because NaOH is very hygroscopic (easy to bind water) so that the H2O contained in the charcoal can react with it [11].

2.4. Preparation of Stock Solution of Cu (II) 1000 ppm

Copper Manufacture mother liquor with a concentration of 1000 ppm is by weighing the solids $CuSO_4.5H_2O$ as much as 3.929 gram using analytical balance and then put into the flask, then add 1000 mL of aquadest up to the mark limit, then shaken until the solid dissolves and is ready to use.

2.5. Determination of optimum mass

The adsorbent mass (cassava peel activated charcoal) used was 300 mg, 400 mg, 500 mg, 600 mg, and 700 mg. Each was put into an Erlenmeyer then added 50 mL of 50 ppm Cu(II) solution, then stirred for 60 minutes using a shaker, then filtered using filter paper, then the adsorbance was measured using an atomic absorption spectrometer [13].

2.6. Determination of the optimum solution pH

50 mL of Cu (II) solution was added to each Erlenmeyer with a concentration of 50 ppm, then the optimum mass of cassava peel adsorbent was added to the Erlenmeyer. Then each Erlenmeyer pH was adjusted with variations of 5, 6, 7, and 9, then the mixture was stirred for 60 minutes using a shaker, then filtered using filter paper and the adsorbance was measured using an atomic absorption spectrometer [13].

2.7. Determination of optimum contact time

50 mL of Cu (II) solution was added to each Erlenmeyer with a concentration of 50 ppm, then the optimum mass of cassava peel adsorbent was added to the Erlenmeyer and adjusted the pH of the solution at the optimum pH (treatment result f). Then each Erlenmeyer was set the stirring time with variations of 15, 30, 60, and 90 minutes, then stirred using a shaker, then filtered using filter paper and then the adsorbance was measured using an atomic absorption spectrometer [13].

2.8. Determination of adsorption capacity with variations in concentration

Cu(II) mother liquor is added to each Erlenmeyer with a concentration of 150, 250, and 300 ppm as much as 50 mL, then the optimum mass of cassava peel adsorbent is added to each Erlenmeyer by setting it at the optimum pH, then stirred using a shaker during the optimum contact time obtained, then filtered using filter paper, then measured the adsorbance using an atomic absorption spectrometer. The Capacity (capacity adsorption) of the cassava peel waste adsorbent was determined by the Langmuir isotherm and the Freundlich isotherm [13].

2.9. Data analysis technique

The data analysis used in this study is the Langmuir isotherm (equation 1) and the Freundlich isotherm (equation 2)

$$\frac{C}{x/m} = \frac{1}{\alpha\beta} + \frac{1}{\alpha}C \quad (1)$$

Information:

$\frac{x}{m}$ = adsorption power of the amount of adsorbate adsorbed per unit mass of adsorbent (mg/g)

C = equilibrium concentration of adsorbate in solution after adsorption (ppm)

α = maximum adsorption power

β = Langmuir constant [14] .

$$log\frac{x}{m} = \log k + \frac{1}{n}\log C \quad (2)$$

Information:

$\frac{x}{m}$ = adsorption power of adsorbate amount adsorbed per unit mass of adsorbent (mg/g)

C = equilibrium concentration of adsorbate in solution after adsorption (ppm)

n = Freundlich's constant

k = power maximum adsorption (mg/g) [14].

3. Results and discussion

3.1. Determination of the Optimum Mass of the Solution for the Adsorption of Cu(II) Ions by Activated Carbon of Cassava Peel Waste

The analysis results concentration of Cu (II) in determining the optimum mass of the solution to the adsorption of metal ions Cu (II) activated carbon of cassava peel waste can be seen in Table 1. Based on Table 1, the relationship curve between adsorbent mass and %Cu (II) adsorbed on cassava peel activated carbon is obtained can bee seen in Figure 1.

Table 1. Results concentration of Cu (II) in determining the optimum mass of the solution to the adsorption of metal ions Cu (II) activated carbon of cassava peel waste

Adsorbent mass (mg)	Abs	pH	Contact time (minutes)	C_i (mg/L)	C_{eq} (mg/L)	C_b (mg/L)	$^x/m$ (mg/g)	% Trapped
300	0,0461	5	60	50	3,127	46,873	7,812	93,746
400	0,0290	5	60	50	1,573	48,427	6,053	96,854
500	0,0242	5	60	50	1,136	48,864	4,886	97,728
600	0,0280	5	60	50	1,482	48,518	4,043	97,036

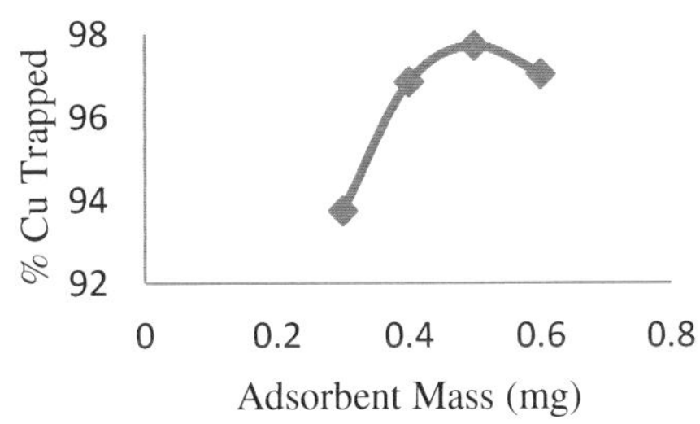

Figure 1. Relationship curve between adsorbent mass and % cu(ii) adsorbed on cassava peel activated carbon

The effect of adsorbent amount is an important parameter because it can determine the capacity of the adsorbent during the addition of adsorbate concentration [15]. In Figure 1 it can be seen that the increase in adsorption that occurs at the adsorbent weight of 300 mg to 500 mg shows that the addition of an increasing number of adsorbents causes the surface of the adsorbent to be wider, so that more adsorbate is adsorbed [16]. The increasing weight of the adsorbent is proportional to the increase in the number of particles and the surface area of the adsorbent, causing the number of binding sites for metal ions to also increase and the absorption efficiency to increase [17].

The decrease in adsorption on the addition of 600 mg adsorbent mass was due to the metal Cu ions having been completely adsorbed. So that the addition of adsorbent mass does not affect the amount of metal ions adsorbed [18]. Under these conditions, the adsorption process was declared to have stopped because based on the value of %Cu that was adsorbed it had approached equilibrium where the number of adsorbate molecules bound to the adsorbent was getting smaller. The situation has reached the saturation point where the surface has been filled with adsorbate [19]. Based on the data in Table 1 and Figure 1, the optimum value for the variation of the mass of the solution using activated carbon of cassava peel as an adsorbent was at a mass of 500 mg which was able to adsorb 97.72% Cu(II) ions. The best adsorbent mass obtained was at a mass of 500 mg. This mass is different from the results of research by Adriansyah [19] who conducted a test of Cu metal adsorption using coffee rind as adsorbent.

3.2. Determination of the Optimum pH of the Solution for the Adsorption of Cu(II) Ions by Activated Carbon of Cassava Peel Waste

The analysis results concentration of Cu (II) in determining the optimum pH of the solution to the adsorption of metal ions Cu (II) activated carbon of cassava peel waste can be seen in Table 2.

The acidity (pH) is a factor that greatly affects the adsorption process of metal ions in solution, because the presence of H^+ ions in solution will compete with cations to bind with active site. In addition, pH will also affect the ionic species present in the solution, so that it will affect the interaction between ions and the active site of the adsorbent [20]. Figure 2 Relationship between pH and % Cu(II) Adsorption on Cassava Peel Activated Carbon.

Table 2. Results concentration of Cu (II) in determining the optimum pH of the solution to the adsorption of metal ions Cu (II) activated carbon of cassava peel waste.

Adsorbent mass (mg)	Abs	pH	Contacts time (Minutes)	C_i (mg/L)	C_{eq} (mg/L)	C_b (mg/L)	x/m (mg/g)	% Trapped
500	0,0218	5	60	50	1,073	48,927	4,892	97,854
500	0,0183	6	60	50	0,754	49,246	4,924	98,492
500	0,0216	7	60	50	1,054	48,946	4,894	97,892
500	0,0617	9	60	50	4,7	45,3	4,53	90,6

At pH 5 the adsorbent has an absorption capacity of 97.85%. The low absorption is due to the absorption of Cu (II) metal ions that do not work optimally in solutions that have an acidic pH value due to the concentration due to the H^+ concentration being too high so that the negative functional group will react with H^+ and prevent the binding of Cu (II) ions [21]. At pH 7 the absorption of the adsorbent decreased by 97.89%. The decrease in absorption at neutral pH because the ions can undergo a hydrolysis reaction in solution, it is not stable in the original ionic form and causes the adsorbent's ability to absorb to decrease [17].

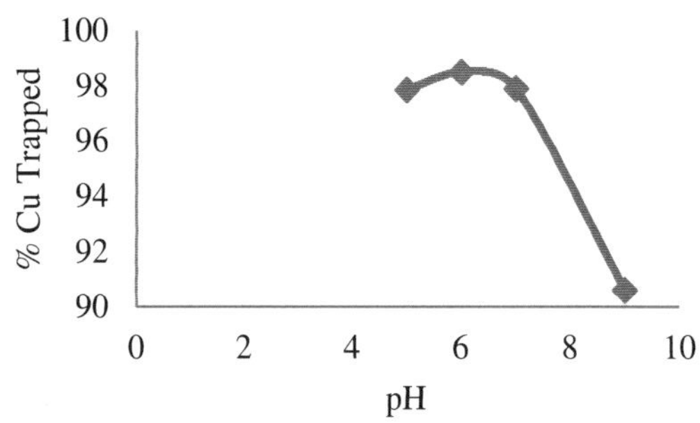

Figure 2. Curve of the relationship between pH and % Cu(II) adsorbed on cassava peel activated carbon

The absorption of the adsorbent at pH 9 decreased significantly, namely 90.6% this is because at pH 9 or alkaline pH, Cu(II) ions will precipitates to form $Cu(OH)_2$ which can prevent adsorption so that Cu(II) metal ions cannot be absorbed properly [22].

$$Cu^{2+}_{(aq)} + 2OH^-_{(l)} \rightarrow Cu(OH)_{2(s)}$$

Based on the above data, the optimum value for the variation of pH solution using cassava peel activated carbon adsorbent is at pH 6 which is able to adsorb as much as 98.49% Cu(II) ions.

The best adsorption pH obtained at pH 6 was different from the results obtained in the Cu metal adsorption test using corncob biomass adsorbent with the best Cu(II) metal adsorption conditions at pH 7 [10] and the optimum pH for Cu ion absorption with coffee husk biomass adsorbent was 4 [23] Based on this, it can be said that the difference in optimum pH is due to the difference in the composition of the biomass used.

3.3. Determination of the Optimum Contact Time of the Solution for the Adsorption of Cu (II) Ions by Activated Carbon of Cassava Peel Waste

The analysis results concentration of Cu (II) in determining the optimum contact time of the solution to the adsorption of metal ions Cu (II) activated carbon of cassava peel waste can be seen in Table 3.

Table 3. Results concentration of Cu (II) in determining the optimum contact time of the solution to the adsorption of metal ions Cu (II) activated carbon of cassava peel waste.

Adsorbent mass (mg)	Abs	pH	Contact time (Minutes)	C_i (mg/L)	C_{eq} (mg/L)	C_b (mg/L)	x/m (mg/g)	% Trapped
500	0,0131	6	15	50	0,628	49,372	4,937	98,744
500	0,0144	6	30	50	0,779	49,221	4,922	98,442
500	0,0127	6	60	50	0,581	49,419	4,941	98,838
500	0,0143	6	90	50	0,767	49,232	4,923	98,466

Based on Table 3, the relationship curve between adsorbent contact time and %Cu (II) adsorbed on cassava peel activated carbon is obtained can bee seen in Figure 3.

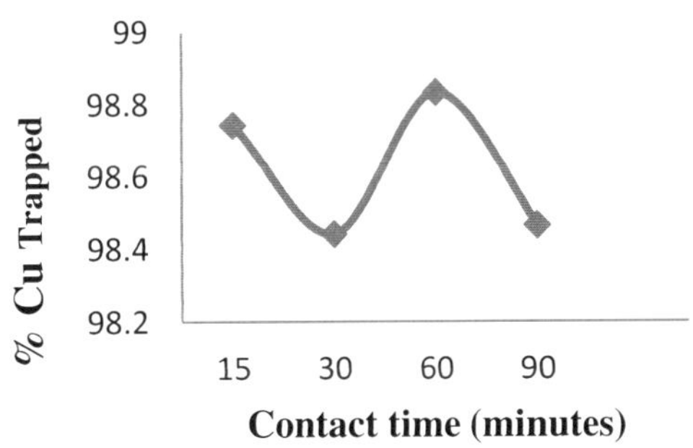

Figure 3. Curve of the Relationship between Contact Time and % Cu(II) Adsorbed on Cassava Peel Activated Carbon

The results of the optimum contact time determination test show that the optimum contact time occurs at 60 minutes with an absorption capacity of 98.84%. Furthermore, in the 15th minute the absorption of Cu metal ions was 98.74% and in the 30th minute the absorption of Cu metal ions decreased by 98.44%. Before reaching the optimum contact time, the adsorption process in the 15th and 30th minutes there was an increase and decrease. This fluctuating value is due to the adsorption that occurs is reversible (back and forth) so that the absorption of the adsorbent becomes unstable on Cu(II) metal ions [24].

After reaching the optimum contact time the adsorption process experienced a decrease in absorption, namely at 90 minutes with an absorption capacity of 99 ,63%, this occurs because of the desorption or re-release of ions that bind to the adsorbent which is experiencing saturation where the pores in the adsorbent are fully filled [25].

The best adsorption time obtained was 60 minutes, which was different from the results obtained in the Cu metal adsorption test using a coffee rind adsorbent with the best condition for Cu(II) metal absorption being 100 minutes [23].

3.4. Determination of Adsorption Capacity with Variation of Cu(II) Metal Ion Concentration by Activated Carbon of Cassava Peel Waste

In this study, the Langmuir adsorption isotherm equation was used to apply to the adsorption of copper ions by activated charcoal of cassava peel. The Langmuir linearity for Cu(II) absorption at various concentrations is shown in Figure 4.

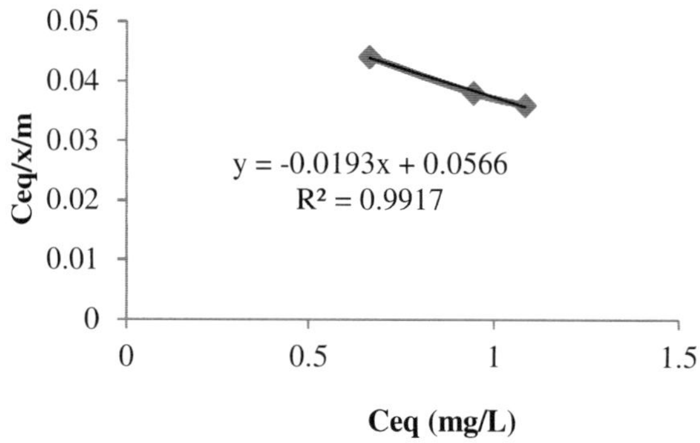

Figure 4. Langmuir Linearity Curve for Cu(II) Absorption at variations in concentration

Langmuir isotherm is used with the assumption that the layer formed is a monolayer layer where the adsorbent bond with the adsorbate is strong enough due to the formation of a chemical bond [26]. Based on the linearity curves for uptake of Cu (II) concentration variation correlation coefficient $R^2 =$ 0.9917 with a maximum adsorption capacity (α) = -51.813 mg/g. In addition, the value of the Langmuir constant (β) is -0.34.

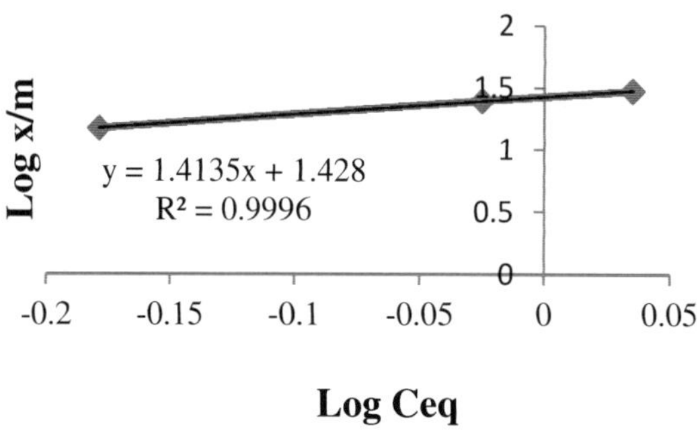

Figure 5. Freundlich Linearity Curve for Cu(II) Absorption at Variations in Concentration

The Freundlich isotherm is used with the assumption that the multilayer layer in which the bond between the adsorbent and the adsorbate occurs due to Van der Waals forces so that the bond is not too strong [26]. Based on the linearity curves for uptake of Cu (II) concentration variation correlation coefficient $R^2 = 0.9996$ with a maximum adsorption capacity (k) of 26.792 L/g, and n value of 6,301 which is a constant Freundlich [27]. The value of n indicates the adsorption characteristics. The adsorption process occurs very well if the n value is 2-10, sufficient if the value is 1-2, and bad if the value is <1 [29]. In this study, the value of n was 6.301 and indicated that the adsorption process was very good.

Adsorption isotherm pattern for each adsorbent mass can be determined by looking at the value of the coefficient of determination (R^2) on the adsorption equation. The largest correlation coefficient value indicates that the adsorption process using the adsorbate is in accordance with the assumed characteristics. From the two graphs above, it can be seen that the adsorption of Cu(II) metal ions by activated carbon of cassava peel waste is more likely to follow the Freundlich isotherm equation than Langmuir, it can be seen from the value of the coefficient of determination R^2,which is 0.9996 (close to number 1). This value assumes that the adsorption that occurs is physical adsorption that forms a multilayer layer and the activated carbon used has a very large surface area so that all Cu(II) metal ion molecules can be adsorbed by forming two or more layers [28].

Conclusion

The conclusions of this study are: that the best conditions (optimal conditions) were obtained with the addition of 0.5 grams of active carbs from cassava peels. The percentage of Cu ion adsorption in these conditions was 97.72%, at pH 6 and a contact time of 60 minutes and the absorption capacity was 98.49%. The maximum adsorption capacity of cassava peel activated carbon to Cu(II) ions at the optimum condition was determined based on the Langmuir and Freundlich isotherm equations. The results obtained were -51.813 mg/g and 26,792 mg/g, respectively.

Acknowledgement

The author would like to thank the Teacher Training Education Faculty of Tadulako University.

References

[1] Imelda D, Khanza A and Wulandari D 2019 Effect of Particle Size and Temperature on the Absorption of Copper (Cu) with Activated Charcoal from Banana Peels (Musa Paradisiaca Formatypica) *J. Technol.* **6** 107–118

[2] Hastuti B, Mudasir M, Siswanta D and Triyono T 2015 Preparation and Pb(II) Adsorption Properties of Crosslinked Pectin-Carboxymethylchitosan Film *Indones. J. Chem.* **15** 248–255

[3] Herwanto B and Santoso E 2006 Pb(II) Metal Ion Adsorption on Crosslinked Chitosan-Cellulose Membrane *Kim. Deed* **2** 9–24

[4] Bird T 1993 *Physical Chemistry for University 2nd Printing* (Jakarta)

[5] Rumhayati B and Retnaningdyah C 2018 Integrative Assessment of Pb and Cd Pollution in Porong Estuaries using Sediment Chemistry, Bioavailability, and Bioconcentration Factor *Indones. J. Chem.* **18** 464–71

[6] Irmawati H, Aprilita N . and Sugiharto E 2018 Adsorpsi Zat Warna Kristal Violet Menggunakan Limbah Kulit Singkong (Manihot esculenta) *BIMIPA* **25** 17–31

[7] Ikawati Z 2010 *Smart About Medicine* (Yogyakarta)

[8] Syahrir I and Syahrir M 2018 The Effectiveness of Used Cooking Oil Purification with Charcoal Adsorbent *Proceedings of the Research Results Seminar* pp 116–120

[9] Dewi Y . and Pratiwi I 2017 Utilization of Cassava Peel Waste (Mahihot Utilissima) in Influencing Fe levels in Water *J. Environ. Sci.* **2** 1–7

[10] Windiastuti R T and Santoso T 2011 Determination of Optimum pH and Contact Time from the Adsorption Process of Cu(II) Ions by Corn Cob (Zea mays) Biomass *J. Akad. Kim.* **9** 224–9

[11] Maulinda L Z N and Sari D 2017 Utilization of Cassava Peel as Raw Material for Activated Carbon *J. Chem. Technol. Unimal* **4** 11

[12] Ariyani, Putri A ., Eka R . and Fathoni R 2017 Utilization of Cassava Peel as Raw Material for Activated Charcoal with Variations in NaOH Concentration and Temperature *Conversion* **6** 7

[13] Pratiwi A . 2016 Efektivitas Pemanfaatan Limbah Tulang Sapi Sebagai Adsorben Logam Cu (Ii) dalam Air (Doctoral dissertation, UII)

[14] Handayani M and Sulistiyono E 2009 Langmuir and Freundlich Equation Test on the Absorption of Chrom (VI) Waste By Zeolite *Proceedings of the National Seminar on Nuclear Science and Technology*

[15] Nurhasni N, Hendrawati H and Saniyyah N 2014 Rice Husk to Absorb Copper and Lead Metal Ions in Wastewater *J. Chem. Val.* **4** 36–44

[16] Irwandi R, Yenti S R and Chairul 2015 Penentuan Massa dan Waktu Kontak Optimum Adsorpsi Karbon Aktif dari Ampas Tebu sebagai Adsorben Logam Berat Pb *J. Chem. Inf. Model.* **52** 1689–99

[17] Harimu L, Rudi L, Haetami A, Santoso G A . and A 2019 Study of Variation of Concentration of NaOH and H2SO4 to Purify Silica from Rice Husk Ash as Adsorbent for Metal Ions Pb2+ and Cu2+ *Indones. J. Chem. res* **6** 81–87

[18] Anjani R . and Koestiari T 2014 Granular Activated Carbon Adsorption used for Adsorbent to

Removal Heavy Metal Pb (II) with Competitor Ion Na+ *UNESA J. Chem.* **3** 159–163

[19] Adriansyah R, Restiasih E . and Meileza N 2018 Biosorption of Heavy Metal Ions Cu(II) and Cr(VI) Using Xantasi Coffee Husk Biosorbent *ALOTROP, J. Educ. Chem.* **2** 114–121

[20] Martina D, Hastuti R and Widodo D S 2016 Peran Adsorben Selulosa Tongkol Jagung (Zea mays) dengan Polivinil Alkohol (PVA) untuk Penyerapan Ion Logam Timbal (Pb2+) *J. Kim. Sains dan Apl.* **19** 77–82

[21] Wahyudianto F . 2016 *Study of Utilization of Blood Clam Shell Waste (Anadara Granosa) as Adsorbent of Pb2+, Cu2+, and Zn2+* (Ten November Institute of Technology)

[22] Syauqiah I, Amalia M and Kartini H . 2011 Analysis of Variations in Time and Stirrer Speed in the Adsorption Process of Heavy Metal Waste with Activated Charcoal *Tech. Info* **12** 11–20

[23] Ginting F 2008 *Testing of a Two-Adsorber Adsorption System Cooler Using 1000 mL Methanol as Refrigerant* (University of Indonesia)

[24] Mohadi R, Saputra A, Hidayati N and Lesbani A 2014 Study of the Interaction of Metal Ions Mn2+ with Cellulose from Sawdust *J. Chem.* **8** 1–8

[25] Masruhin, Rasyid R and Yani S 2018 Heavy Metal Sorption of Lead (Pb) using Lignin Isolated from Rice Straw *J. Chem. Process Eng.* **3** 6

[26] Sulastri S, Nuryono, Kartini I and Kunarti E . 2014 Kinetics and Balance of Adsorption of Chromium (iii) Ions in Solution on Silica Compounds and Modified Silica Synthesized from Siti Rice Husk Ash *Sci. Res. J.* **19** 33

[27] Almufarij R 2014 Removal of Crystal Violet Dye from Aqueous Solutions Onto Date Palm Leaf Without the Sharp Spines: Adsorption and Kinetic Studies. *J. Am. Sci.* **9** 311–315

[28] Suhaeri A, Maryono and Side S 2014 The Adsorption Capacity of Cassava Peel Activated Charcoal Against Crions6+ *J. Chem.* **15** 95–104

Journal of Physics: Conference Series **2126** (2021) 012023 doi:10.1088/1742-6596/2126/1/012023

What we can learn from 2018 liquefaction in Central Sulawesi: Stories from the survivors

U Wahyono[1*], and N M Wiwik Astuti[2]

[1]Sub-Department of Physics Education, Faculty of Teacher Training and Education, Universitas Tadulako, Jl. Soekarno Hatta KM. 9 Kampus Bumi Tadulako Tondo Palu-Central Sulawesi, Indonesia.

Corresponding author: uwahyono@gmail.com

Abstract. This paper provides the stories of the liquefaction survivors about their experiences during and after the disaster. It is a phenomenological study that used the semi-structured interview to obtain the data. The data were acquired from 30 participants. There are some important points that can be summarised from the stories of the survivors. During the disaster, local people inclined to underestimate the quake at the beginning of the shaking. It could indicate the level of their risk perception and awareness on disaster. In addition, a unique finding from this study was that the survivors had conducted the initial intervention priority as basic resources for survival such as shelter and information of the victims' identity. Also, regarding the post-disaster data, the results indicated that the survivors experienced the symptoms of trauma due to the disaster. It indicates that the psychological aspect of the survivors was neglected during the recovery process. On that regard, long-term traumatic recovery activities should be put into accounted during the recovery processes for all the victims of a disaster regardless of their ages and genders. Another implication of this study is the opportunity to use these stories into learning activities, both in learning about science or disaster mitigation.

1. Introduction

In 2018, a major earthquake with a magnitude of 7.4 Mw struck Palu, the capital city of Central Sulawesi, along with the surrounding regencies of Donggala, Parigi Moutong and Sigi. The strong ground shaking triggered other disasters such as tsunami, liquefaction and landslide in different parts of the region. While the intense quake destroyed buildings such as public facilities and houses, the tsunami destroyed the coastal areas. In Southern Palu, soil liquefaction buried the whole villages and a landslide closed the main access of Dolo to the outside which isolated the survivors for a couple of weeks.

People in Sulawesi have been familiar with an earthquake along with the tsunami and landslide. Due to the geographical situation of Indonesia, the locals have witnessed these events frequently through mass media or by direct experiences. However, liquefaction had left the survivors with the most disbelief. Although the earthquake alone had caused tremendous damages, the trace of liquefaction was insanely horrifying to the affected areas. In the Central Sulawesi case, the liquefaction happened owing to the soil structure which is composed of thick alluvial deposits and contains silt and clay, and it has shallow groundwater that then promoted the flowability of the liquefied soil [1–3].

Months after the event, liquefaction and its catastrophic effects had attracted the attention of engineers and seismologists to learn about the soil profiles in Central Sulawesi. Numerous research had been conducted to measure and investigate the Palu-Koro fault system and the causes of that hazardous disaster (e.g. [4–6]). The other studied the methodology for constructing house and building and groundwater control to reduce the risk of liquefaction [7,8]. However, the extent of possible chronologies of the event and resulting community impacts are not well understood or well documented.

There is a dearth of understanding about the processes people go through the deadly events. The scarceness of documents of survivors' stories indicated the lack of space currently given to talk and share their experiences, and to be heard. Listening to the actual stories of the survivors is imperative especially in terms of the post-disaster recovery process [9].

After the disaster, we learned that the local people defined liquefaction as Nalodo, which means collapsed in the mud. The existence of a local term for liquefaction denotes that liquefaction had occurred before. However, the information resources about this phenomenon, such as notes and studies are scarcely reached by the public. This study aims to document the stories of survivors during the disasters and after the disaster to prevent their stories being erased by time. The narrative of the survivors can be used to unfold details that cannot be seen through scientific theories. In addition, the existence of the documents containing survivors' stories can depict what they had been through during the disaster. Deepening understanding of what had happened in people's lives can be used as a means to gauge the level of resilience and vulnerability of the survivors in the aftermath of the disaster. Furthermore, it can be utilized as a source for emotional and political investigation and as a base for formulating the proper approach for post-disaster trauma recovery. To be more specific, this study aims to provide an overview of the liquefaction experiences from the perspectives of the eyewitnesses, during and after they survived from the disaster, including the descriptions of soil liquefaction, the survival attempts from mud trap and the post-disaster life.

2. Method

In this study, we focus on survivors' experience during the liquefaction disaster, henceforth we used a phenomenological methodology. The use of phenomenology as a study of phenomena is to describe how things appear to people through experiences and to provide a detailed explanation of people's real life experiences [10].

Participants in this study were the survivors of the earthquake on 28 September 2018, who had experienced the direct soil liquefaction event. We used purposive sampling to identify the primary participants based on our judgment and the purpose of the research. In order to trace additional participants, we used snowball sampling by asking one participant to recommend others for interviewing. There were 30 participants, 16 were women, 14 were men. Their ages ranged from 40 to 55 years olds.

The interviews were conducted by the research team and two volunteers from the affected areas. Before the interview, we gave the guide of the interview to them and highlighted the information that should be gained from the participants. We intended to make the interviews as conversations that were loosely structured to allow the participants to tell their experiences. We used semi-structured interviews which began by asking the participants "Can you tell me about the event, from the moment before the event to the moment when you felt that you had been saved?" The follow-ups questions were based on the initial stories of the participants.

Prior to the interview, the interviewers explained the purposes of the interview to participants and they read the consent forms. Interviews were conducted in places that were chosen by the interviewees which were mostly in their houses. We interviewed participants face to face approximately 22 months after the disaster. The length of the audio-taped interviews ranged from 4 to 30 minutes. We used pseudonyms in lieu of participants' names to protect confidentiality.

Audio-taped were transcribed before being analyzed for codes and categories. The constant comparative method was used to analyze data by assigning codes by sorting the data into groups of categories that should reflect the purpose of the study [11]. Then, the group of categories translated from Bahasa Indonesia into English by the second author.

3. Result and Discussion

3.1. During the disaster

3.1.1 First Responses to the Quake

Most of the survivors indicated sharing the common first responses to the quake, such as running to the open space. However, some underestimated the shaking at the beginning. They stated that experiencing frequent earthquakes made them dismissive of the quake. As Jamal expressed:

> Before the disaster, my family and I were at the house. After the first shock at around 4 pm, we have been used to earthquakes like that because we are in Palu. We can say that this (the quake) happens very often here so we were still at home calmly like usual. And then we were shocked, we were shaken very intensely then we left our house.

Some survivors perceived the mainshock would be a low-level earthquake thus they intended to stay inside their homes before going outside. These frequent low-intensity shaking led people to ignore the disaster before the mainshock occurred. This finding supports the previous studies that reported how experiencing the frequent and minor effects of small quakes can lead people to decrease their awareness and preparedness toward disaster [12–14].

3.1.2. Confronting the Soil Liquefaction

The time interval between the quake and liquefaction was short. The liquefaction occurred right after the quake stopped. Seven participants mentioned that before the mud came, they saw black-thing came from the upper areas (eastern area) along with a strange loud sound. For instance, Azis said:

> Suddenly, I heard a roaring sound...not long after I heard that sound, I looked towards the upper area (eastern part of Jono Oge) and I saw there were black waves. I did not know that that was mud.

Moreover, participants likened the devastating experience of being trapped in the current of mud as being in the blender machine. For example, a participant explained:

> When being dragged, I was twisted, sank, hit wood, and was crammed between them... the mud was like hard, soft and hot... and wavy, like sea waves...My mind was very chaotic, I was dragged by mud, I felt like my body was twirled in mud, do you know how fruits in a blender machine? Yes, that was how my condition at that time, all of my body was covered by mud. I could not stop or direct my body when I was dragged by it because it was so strong.

The location of Petobo and Jono Oge districts was less than 10 km from the epicenter. This close distance to the epicenter might result in a strong sound, like roaring and thunder sound, heard by the participants. Although the sound does not contribute to the estimations of macroseismic intensity, it caused massive fear and panic to survivors [15]. Some participants also experienced lateral spreading ground failure when they pointed out how they were running at the same place because the road they were stepping on was also moving. The road remained the same in shape, but they moved as the result of liquefied soil under the surficial soil layers. They also saw the shift up and down movement of their houses while moving. It indicates that all types of ground failure due to liquefaction occurred. In addition, the hot temperature of mud indicated the possibility of soil boiling.

From the stories, there were some science concepts found, such as sound, substance characteristics, and types of ground failure. Since the disaster curriculum currently has gained the attention of the central and local government in Indonesia, including the actual incident stories in the disaster or science curriculum could give positive influence for students. It is in accordance with the statement of Saleh and Pendley [16] and Rae [17] who believed that a safety or disaster curriculum should involve the real cases or experiences of the disaster and accidents.

3.1.3. Escaping from Liquefaction

The participants encountered different circumstances and situations during their attempt to stay alive amidst the current of mud. They told stories about their efforts to survive while they were dragging and buried by the mud; what they had been through and the states of their mind while waiting for the fortunes and miracle to come. The length of hours of survivors trapped in the mud before getting out of or being rescued from the mud was varied, ranging from 1 hour to more than 10 hours.

While swayed away and surrounded by the mud, participants indicated the importance of the existence of hard and solid objects to hold. They helped the survivors to stay on the surface. Nusa labeled the solid objects as the treasure. He claimed that most of the survivors were those who were capable of climbing to the roofs of the sinking houses. Other objects that the participants used to help them were above the surface roots and the stems of the fallen trees, the cable of power poles and timbers.

It shows that it is imperative to look for a light yet solid material for footing while escaping from the earthquake.

In the middle of confusion due to the unprecedented and horrific situation, most of the participants expressed their reliance on and surrender to God. When they were asked how they could survive, most of the participants expressed that they begged the almighty for help. In addition, they claimed that surviving from that terrific event was owing to God's help and that that time was not yet their *ajal* or the time to die. Besides, survivors also stressed the imperative of being mindful and focused on that situation. Although previous studies have reported the significance of religion for mental problems after the disaster (e.g. [18–20]), there has been no study found which specifically researches what and how is the role of belief and religion for the survivors during a disaster. It has not been justified, however, turning to God might help the survivors to stay calm in a chaotic situation. In the liquefaction case, they claimed that focus and mindfulness helped them to mind what they can hold onto and what they can step on. Running hastily without a destination will increase the risk rate for injury [21]. Further, self- and mutual aids are significant in the early time of rescue after the disaster.

3.2. Peri-Disaster

3.2.1. First quick response after the disaster

After surviving from the mud, most of them reported that they were escorted to or were ordered to gather at the mosques in a neighboring village. When they arrived, some people wrote their identity and built emergency tents with available materials. A participant reported:

> When we (the survivors) were taken to the mosque, our names were immediately recorded, to make it easier for the family to find a family that was washed away by the mud, then my son came to find me.

Participants have shown that they had done the initial intervention priority as basic resources for survival such as shelter and information of the victims' identity [22]. It was a unique finding because socialization related to disaster mitigation is scarcely conducted for locals. Disaster education even disseminates very slowly in Central Sulawesi. It is interesting how, albeit a dearth of information about the first aids, the locals have had the initiative to provide the refugee camp and the information of the victim. The residents in Central Sulawesi had experienced multiple earthquakes. It needs further investigation whether these first quick responses were promoted by their disaster subculture due to the frequent disastrous hazards that had affected them.

3.3. Post-disaster: The Trauma

Due to the deadly and near-death experience, the participants indicated the post-disaster trauma by remembering the event. A participant said that she even could not stand the sound of the wind. She will immediately feel scared and uneasy if she hears disaster news, such as floods, and the sound generated by the wind. She recalled:

> It's hard to tell (about it). The wound reappeared. I remembered my family. If I tell you the story, what appears in our mind is the destruction of the Petobo (her village). It's like a wound, it hurts my heart. When I told you this story, I got goosebumps. Alhamdulilah, all of my family survived. But since we (the neighbours) were from the same neighbourhood, the same tribe, we were one family. It is painful.

It indicates that the participants' feelings are still trapped in the situation. The memories of losing family members, loved ones, and their properties put the victims in a state of despair and shock. Moreover, survivors of liquefaction suffered from losing their neighbourhood where they had spent their entire life. The neighbourhood is associated with both human and social capital. Losing it was like losing the entire community, the families and all the memories of their lives. A sense of losing a place was the most catastrophic impact of natural disasters on the survivors [23]. The liquefaction forced the survivors to reside in a new environment and community. They experienced the environmental change, the loss of resources, daily routine, social support which connect with stressors [22,24]

This study does not highlight the psychological effect of disaster experience. However, a glimpse of the survivors' answers supports broadening concepts regarding the nature of disaster impact on acute stressful experiences (e.g. [25–27]). It demonstrates how the survivors fail to recover from that horrific event. It indicates that many survivors may experience trauma after perceiving a mortal risk to themselves at the event. This issue needs to be addressed since pre-existing trauma could risk developing psychopathology after a disaster [27,28]. Also, injured survivors of the previous earthquake would likely get injured again in future events due to the trauma [21]. Moreover, long-term psychological distress could be experienced by the survivors especially by those who had severe traumatic experiences [24]. Accordingly, survivors should get support in terms of mental aids to recover from this trauma.

4. Conclusion

The geographical situation of Indonesia leads to the frequent earthquakes felt across the country. However, these recurrent events have caused local people to underestimate the quake on 28th September 2018. Instead of increasing the awareness and preparedness of disaster, disaster experiences with an insignificant impact on them and their surroundings seemed to generate their ignorance towards the possibility of immediate risk from a disaster. The collected data of this study indicated that all types of ground failure; flow failure, ground oscillation and lateral spreading, occurred in this event. Further, since most of the survivors mentioned God in their stories, it is intriguing to find the role of their faith in God during the disaster, such as how they feel about the disaster and the process of their survival. The results could benefit on how to build resilience for mental recovery. Moreover, although this study was not weighed in the psychological aspect, it is worth considering for future work. The high frequency of rejection from survivors to participate could be a sign of posttraumatic stress disorder (PTSD). Also, there are information from the participants regarding the occurrence of the liquefaction that can be involved in science and disaster curriculum. We are contemplating whether it would be possible to include the survivors' stories into classroom activities as a method for disaster risk reduction (DRR) education dissemination. Bringing more contextual topics to the classroom would certainly increase students' engagement. However, it requires further investigation whether it will trigger students' interest in learning, or trigger their nightmare.

Acknowledgement
We highly acknowledge the support of the Program Kompetisi Kampus Merdeka (PK-KM) Grant of the Education, Culture, Research, and Technology of the Ministry of the Republic of Indonesia, with funding identity : 119/E1/KM.05.03/2021.

References

[1] Putra R R, Kiyono J, Yoshimoto Y, Ono Y and Syahril 2016 Determined soil characteristic of Palu in Indonesia by using microtremor observation *Int. J. GEOMATE* **10** 1737–42
[2] Sassa S and Takagawa T 2019 Liquefied gravity flow-induced tsunami : first evidence and comparison from the 2018 Indonesia Sulawesi earthquake and tsunami disasters *Landslides (2019)* **16** 195–200
[3] Widyaningrum R 2012 *Penyelidikan geologi teknik potensi liquifaksi daerah Palu , Provinsi Sulawesi Tengah* (Bandung: Badan Geologi)
[4] Abdullah A I and Abdullah 2020 A field survey for the rupture continuity of Palu-Koro fault after Donggala earthquake on September 28th, 2018 *J. Phys. Conf. Ser.* **1434**
[5] Fang J, Xu C, Wen Y, Wang S, Xu G, Zhao Y and Yi L 2019 The 2018 Mw 7.5 Palu earthquake: A supershear rupture event constrained by InSAR and broadband regional seismograms *Remote Sens.* **11** 1–15
[6] Jaya A, Nishikawa O and Jumadil S 2019 Distribution and morphology of the surface ruptures of the 2018 Donggala–Palu earthquake, Central Sulawesi, Indonesia *Earth, Planets Sp.* **71** 1–13
[7] Adalier K and Elgamal A 2004 Mitigation of liquefaction and associated ground deformations by stone columns *Eng. Geol.* **72** 275–91
[8] Maithili K L 2017 A discussion of liquefaction mitigation methods *Int Res J Eng Technol* **4** 1830–3
[9] Kargillis C and Kako M 2014 Disaster survivors: A narrative approach towards emotional recovery *Aust. J. Emerg. Manag.* **29** 25–30

[10] Finlay L 2009 Debating phenomenological research methods *Phenomenol. Pract.* **3** 6–25
[11] Merriam S B 1988 *Qualitative research and case study applications in education.* (San Fransisco: Jossey-Bass)
[12] Paton D, Mamula-Seadon L and Selway K L 2013 *Community resilience in Christchurch: Adaptive responses and capacities during earthquake recovery*
[13] Simpson-Housley P and Curtis F A 1983 Earthquake occurrence, experience and appraisal in Wellington, New Zealand *Prof. Geogr.* **35** 462–7
[14] Paton D, Johnston D, Mamula-seadon L and Kenney C M 2014 Recovery and Development: Perspectives from New Zealand and Australia *Disaster and Development:Examining Global Issues and Cases* (New York: Springer)
[15] Mäntyniemi P 2004 Pre-instrumental earthquakes in a low-seismicity region: A reinvestigation of the macroseismic data for the 16 November 1931 events in Central Finland using statistical analysis *J. Seismol.* **8** 71–90
[16] Saleh J H and Pendley C C 2012 From learning from accidents to teaching about accident causation and prevention: Multidisciplinary education and safety literacy for all engineering students *Reliab. Eng. Syst. Saf.* **99** 105–13
[17] Rae A 2016 Tales of disaster: the role of accident storytelling in safety teaching *Cogn. Technol. Work* **18** 1–10
[18] Anastasova R A 2014 *Understanding the Role of Religion in Coping after Trauma : Resilience, Post-traumatic Growth and Difference in Coping Mechanisms* (Tilburg UNiversity)
[19] Henslee A M, Coffey S F, Schumacher J A, Tracy M, Norris F and Galea S 2015 Religious Coping and Psychological and Behavioral Adjustment after Hurricane Katrina *J Psychol* **149** 630–642
[20] Shamsalinia A, Ghaffari F, Dehghan-Nayeri N and Poortaghi S 2017 The Life Process of Children Who Survived the Manjil Earthquake: A Decaying or Renewing Process. *PLoS Curr.* **9** 0–18
[21] Kang P, Lv Y, Deng Q, Liu Y, Zhang Y, Liu X and Zhang L 2017 *Investigating lushan earthquake victims' individual behavior response and rescue organization* vol 14
[22] Freedy J R, Shaw D L, Jarrell M P and Masters C R 1992 Towards an understanding of the psychological impact of natural disasters: An application of the conservation resources stress model *J. Trauma. Stress* **5** 441–54
[23] Diaz P and Dayal P A 2008 Sense of place: A model for Community Based Psychosocial Support Programs *Australas. J. Disaster Trauma Stud.* **1** 1174-4707.
[24] Tanaka E, Tennichi H, Kameoka S and Kato H 2019 Long-term psychological recovery process and its associated factors among survivors of the Great Hanshin-Awaji Earthquake in Japan: A qualitative study *BMJ Open* **9** 1–8
[25] Jenkins, R and Meltzer H 2012 *The Mental Health Impacts of Disasters*
[26] Makwana N 2019 Disaster and its impact on mental health: A narrative review *J Fam. Med Prim Care* **8** 3090–3095
[27] Tang B, Liu X, Liu Y, Xue C and Zhang L 2014 A meta-analysis of risk factors for depression in adults and children after natural disasters *BMC Public Health* **14** 1–12
[28] Bromet E J, Atwoli L, Kawakami N, Navarro-Mateu F, Piotrowski P, King A J, Aguilar-Gaxiola S, Alonso J, Bunting B, Demyttenaere K, Florescu S, De Girolamo G, Gluzman S, Haro J M, De Jonge P, Karam E G, Lee S, Kovess-Masfety V, Medina-Mora M E, Mneimneh Z, Pennell B E, Posada-Villa J, Salmerón D, Takeshima T and Kessler R C 2017 Post-traumatic stress disorder associated with natural and human-made disasters in the World Mental Health Surveys *Psychol. Med.* **47** 227–41

Characterization of patchouli oil (*Pogostemon cablin* Benth) production of Tinombala Village, Ongka Malino District, Parigi Moutong Regency

P H Abram[1], W Putri[1], S Nuryanti[1] and S M Sabang[1]

[1]Chemistry study program, mathematics and sciences division, faculty of teacher training and education, Tadulako University, Jl. Soekarno Hatta Km 9 Palu 94118 Indonesia.

*Corresponding author: paulus_hengky@untad.ac.id

Abstract. Patchouli oil comes from Tinombala village which has never been tested for the Indonesian National Standard (SNI). The purpose of this study was to determine the quality of patchouli oil by characterizing parameters of physical and chemical properties including color, refractive index, specific gravity, solubility in alcohol, acid number, ester number, iron content (Fe), and patchouli alcohol and alpha copaene content. . Sampling was done by random sampling. The results show a specific gravity of 25/25oC sample A = 0.9564, B = 0.9658, and C = 0.9590, refractive index (nD20) sample A = 1.5111, B = 1.5085 and C = 1.5086, the solubility in alcohol using ethanol of the three samples is >1:10 while using methanol 1:10, the acid number of the sample A=21.0375, B=9.1162 and C=19.6350, the ester number of the sample A=28 ,05, B=21.04 and C=21.04, iron (Fe) content of sample A=1.84210 mg/L, B=0.186842 mg/L and C=6.78950 mg/L, patchouli content alcohol sample A=,16,96%, B=15,52% and C=14,55% and for alpha copaene not detected. The results showed that the quality of patchouli oil in Tinombala village did not meet the Indonesian National Standard (SNI). Only the specific gravity, refractive index, and iron content are by the quality requirements, while the acid number and ester number and the main component determining the quality of patchouli oil, namely the patchouli alcohol content are still below the Indonesian National Standard, which is below 30%.

1. Introduction

Patchouli oil production in Indonesia is one of the largest foreign exchange earners when compared to other essential oils. About 80% of the world's patchouli oil demand is provided by Indonesia [1]. So that the development of patchouli cultivation as a staple for patchouli oil production, has a very good opportunity and potential to help increase farmers' income if it meets its quality standards, it is known that the demand for patchouli oil in the territory of Indonesia continues to increase along with the increasing demand for patchouli oil in the world market. . In addition, the prospects for the patchouli oil market in the country are also getting better and have quite a promising world market opportunities [2]. This is driven by the growing demand in various aspects such as the cosmetic and fragrance industry [3].

Patchouli oil has many uses including being used as an important raw material in the chemical field and becoming a much sought after commodity because it contains a chemical compound, namely patchouli oil. The main benefit of patchouli oil is as a fixative or binder in various industries, including in the fields of soap, perfume, cosmetics, aromatherapy because it has a distinctive odor and is often used as a mixture for making drugs [4]. In addition, it is widely used as raw material and admixture in antiseptics, perfumes, pharmaceuticals, chemicals, insecticides, and aromatherapy treatments [5]. Patchouli oil is also commonly used for preserving goods, food, giving aroma household products, as a mixture, toothpaste, deodorant, lotion as well as other basic industrial needs [6].

According to [3] patchouli cultivation in Indonesia has spread widely in several areas including Aceh, North Sumatra, West Sumatra, Bengkulu, South Sumatra, West Java, Central Java, and East Java. Some areas also began to develop patchouli such as South Sulawesi, East Kalimantan, Central Kalimantan and Central Sulawesi. With the largest distribution of this plant, automatically in Indonesia there are many patchouli oil processing industries, both small and medium scale industries to support market demand, although the demand for patchouli oil continues to increase along with the increasing demand for patchouli oil in the world market, but in terms of quality. Patchouli oil produced in Indonesia has not changed much due to many factors and constraints in the industry such as there are still many farmers who do not know the quality standards of patchouli oil, low productivity, yields and oil quality that are still below standard, patchouli oil processing techniques are still relatively simple, in addition to the low quality of plant seeds, cultivation methods that are still simple and vary from region to region, the development of various diseases, using simple equipment and handling of production results that have not been maximized so that this affects the low selling price of patchouli oil in Indonesia. Patchouli oil with better quality has a higher price when compared to patchouli oil with poor quality, the main component of patchouli oil that affects the quality of patchouli oil is the high content of patchouli alcohol ($C_{15}H_{26}O$) in patchouli oil meaning that the better the characteristics of patchouli oil produced.

According to Rahman et al. [7], Central Sulawesi Province is one of the areas that in recent years has become interested in developing patchouli commodity and has become one of the leading crops for farmers because it has a fairly good selling value. Based on data from Indonesian Plantation Statistics in 2018 in Central Sulawesi, the patchouli plant area reached 1,584 ha with the total productivity of patchouli reaching 257 kg/ha. Central Sulawesi has the largest planted area on the island of Sulawesi, when compared to other Sulawesi provinces. Parigi Moutong Regency is one of the regencies in Central Sulawesi that produces patchouli oil, precisely in Ongka Malino District. Based on data from Indonesian Plantation Statistics in 2016 in Parigi Moutong Regency, the total productivity of patchouli plants can reach 400 kg/Ha [8]. This is evidenced by the presence of a large number of patchouli oil refining home factories in Tinombala village and almost all farmers grow patchouli. However, the patchouli oil processing process developed by patchouli farmers in Tinombala village still applies a simple cultivation technology, namely the Boiled System method (Water Distillation) using an iron plate which is modified again according to the needs of distillation, farmers do not pay attention to the quality of the tools used in the distillation process. , many farmers have limitations in the field of knowledge of patchouli oil refining, farmers do not know the basic characteristics of patchouli oil which is an important subject of oil that can be traded at relatively expensive prices, farmers have never been given counseling by the Oil Supervisory Agency. In addition, the low production and high costs of refining patchouli oil and the processing process are still far from SNI processing standards and the price of patchouli oil in Tinombala village is still relatively cheap when compared to patchouli oil on the market.

Conducted a study on the effect of fractional distillation temperature range on patchouli alcohol content in patchouli oil [3]. Conducted a study on the effect of the fractional distillation on an increment of patchouli alcohol content in patchouli oil [9]. Conducted research on Applying system dynamics for predicting the strengths, weaknesses, opportunities, and treats of patchouli oil agroindustry in West Sumatra, based on the simulation results, there are six strengths, six weaknesses, two opportunities, and three threats of patchouli oil agroindustry forces in West Sumatra The differences in this study are based on location, time and conditions which are known to vary from region to region [10]. Although many have described this type of research, this location is the type of research that is needed to find out initial information about the quality of patchouli oil produced. This is because patchouli oil refining still uses traditional methods and there are many productivity problems.

So that special attention is needed to increase the production of patchouli distillation in the village with the initial step of conducting a scientific study on the purity and quality of patchouli oil with the aim of the study being to determine the main characteristics of patchouli oil produced by the village of Tinombala, because previously there was no information that could describe the actual quality. From patchouli oil, which is the basic capital as a reference for determining the economic price of patchouli oil in the national and world markets [7].

2. Materials and Methods

2.1 The Tools and materials

Equipment used in this study included a pycnometer, 100 mL measuring cup, 50 mL beaker, 50 mL erlenmeyer, refractometer, polarimeter, boiling stone, electric bath, analytical balance, a set of 990tools. Atomic AbsorptionSpectrophotometer (AAS) PG Instruments, a set oftools for Gas Chromatography Mass Spectrometry (GC-MS)gas chromatography-mass spectrometry (GC-MS) brand Thermo Scientific Trace 1310 Gas Chromatography-ISQ 700 Mass Spectrometer.

The chemicals used include ethanol, methanol, diethyl ether, KOH, HCl and phenolphthalein indicator solution, ferroammonium sulfate, acetone, silver nitrate ($AgNO_3$), iron (III) chloride tetrahydrate ($FeCl_2.4H_2O$), sodium chloride (NaCl), chloroform and nitric acid (HNO_3).

2.2 Data Collection

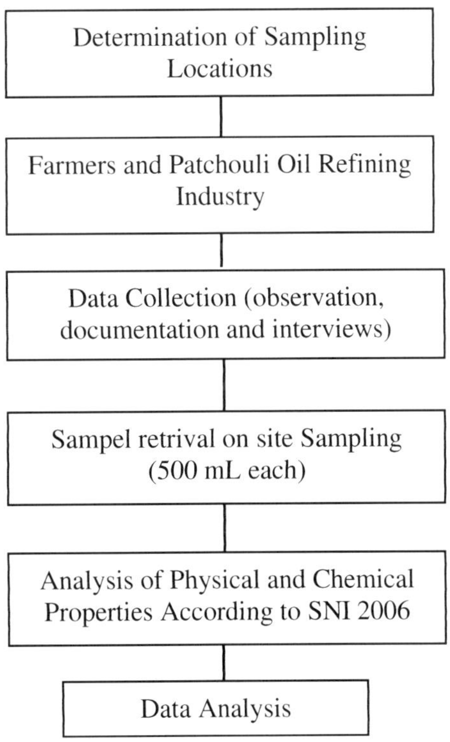

Figure 1. Collection scheme for patchouli oil refining industry players and patchouli oil sampling

2.3Data analysis

Data analysis was based on the 2006 Indonesian National Standard [11].

2.3.1 The color

20 mL of patchouli oil sample was put into a test tube, the test tube was propped up on white paper or cardboard, then the color of the oil was observed with an eye distance of ± 30 cm.

2.3.2 Specific Gravity

Weigh the empty pycnometer (m), fill the pycnometer with distilled water. Dip the pycnometer into a water bath at a temperature of 25°C ± 2° C for 30 minutes, then weigh the contents (m_1). Empty the pycnometer, washed and cleaned, then the pycnometer is filled with patchouli oil samples, then put in the bath under the same conditions.

2.3.3 Refractive Index

Refractometer at a reading temperature of 25°C ± 2°C. Before the oil sample is placed in the tool, the refractometer prism is cleaned, then the patchouli oil sample is dripped on the prism, the prism is closed and then adjust the slide until a clear boundary line is obtained (dark and light). Further readings are made when the temperature is stable.

2.3.4 Solubility in alcohol (ethanol and methanol)

1 mL of patchouli oil sample was put into a 25 mL measuring cup, add absolute alcohol (ethanol and methanol) dropwise. Shake after each addition until a clear solution is obtained. The absolute volume of alcohol added to the sample was recorded until the solution was clear.

2.3.5 Acid Number

Weigh 4 g ± 0.05 g of patchouli oil sample into an Erlenmeyer, then dissolved in 5 mL of ethanol, added 5 drops of phenolphthalein indicator. Then titrated with 0.1 N KOH until the solution turns pink.

2.3.6 Esters Number

Before testing the sample, a blank test was performed using ethanol. After that, weigh the 4 g patchouli oil sample ± 0.05 g then put it into an Erlenmeyer, added 25 mL of 0.5 N KOH solution in alcohol and boiling stone, then refluxed on a water bath for 1 hour after which the solution was allowed to stand. Add 5 drops of phenolphthalein solution and titrate with 0.5 N HCl until a color change is obtained.

2.3.7 Determinants of Iron (Fe), Patchouli Alcohol and Alpha Copaene

Contentscontent the iron (Fe) of patchouli oil samples can be determined using the AAS tool by comparing the absorbance value of patchouli oil samples with the standard solution curve. Meanwhile, the patchouli alcohol and alpha copaene levels of the oil samples were determined based on the analysis of the relative area of the gas chromatogram as measured by the GC-MS instrument by comparing the peak area with the total area of all detected components.

3. Results and Discussion

The results of the analysis of the physicochemical properties of the oil patchouli obtained from theindustry Tinombala distillery in the village community (samples A, B, and C) are presented in Table 1. Patchouli oil sampling in Tinombala village, Ongka Malino district, Parigi Moutong district, using a random sampling technique by taking patchouli oil samples at 3 refining industries from 10 patchouli oil refining industries in Tinombala village. The sampling technique using this method provides equal opportunities for each member of the sample to become the research sample. This is done by considering the basic ingredients of the kettle or distillation equipment used by the community, namely iron plate material and the method used in 10 industrial distilleries using the same distillation method, namely the water distillation system.

Table 1. Results of quality analysis of patchouli oil production in Tinombala Village, Ongka Malino District, Parigi Moutong Regency.

Characteristics/ Parameters of Observation of	SNI Quality Requirements for 2006	Patchouli Oil Samples in Tinombala Village			Average
		A	B	C	
Color	Yellow-brown light reddish	brown reddish	yellow	brown more reddish	
Specific Gravity (25°C)/ (g/L)	0.950-0,975	0.9564	0.9658	0.9548	0.9590
Refractive Index (nD^{20})	1.507-1.515	1.5111	1.5085	1.5086	1.5094
Solubility In Alcohol (Ethanol and Methanol)/(mL)	Clear solution or light opalescence in volume ratio 1:10	Ethanol =1:15 Methanol =1:9	Ethanol =1:12 Methanol =1:10	Ethanol =1:18 Methanol =1:10.5	Ethanol =1:15 Methanol =1:9.8
Acid Number	Maximum8.0	21.0375	9.1162	19.6350	16.5962
Esters Number	Maximum20.0	28.05	21.04	21.04	23.38
Iron (Fe) /(g/mL)	Maximum 25	1.84210	0.186842	6.78950	2.93948
Alpha Copaene Content (%)	Maximum 0.5	-	-	-	-
Patchoul Alcohol Content (%)	Minimum 30%	16.96	15.52	14.55	15.67

3.1 Color

Color is the first visual attraction that can state that patchouli oil has good quality, the results of patchouli oil color testing by visually visualizing the three samples ie patchouli oil sample A is reddish- brown, sample B is yellow and sample C is more reddish-brown when compared to sample A. So in general, when compared with the Indonesian National Standard, the three patchouli oil samples are following the standard parameter specifications. The color intensity of patchouli oil is also determined by the amount of color pigment contained in it. In addition, storage time and storage place also affect the color of patchouli oil. The storage area used in the patchouli oil refining industry in Tinombala village uses used jerry cans made of plastic [12].

3.2 Specific Gravity

Gravity The specific value can be determined by comparing the weight of patchouli oil with water (aquades) using the same volume and temperature (25°C). The specific gravity of patchouli oil according to the standard ranges from 0.950 to 0.975 g/mL [3]. The results in Table 1 above show that the specific gravity of the three patchouli oil samples (samples A, B, and C) has met the SNI standard. Sample A was 0.9564 g/mL, sample B was 0.9658 g/mL and sample C was 0.9548 g/mL. The density of patchouli oil is not influenced by the volume of patchouli oil itself but the mass of the chemical components contained in patchouli oil itself such as patchouli alcohol and other chemical compounds. In addition, the content of water, fat and minerals in patchouli oil can also affect the specific gravity of patchouli oil. So that the specific gravity of patchouli oil cannot be used as the only measure of the quality of patchouli oil is said to be good. Therefore, it can be said that each mass of chemical components contained in patchouli oil such as sesquiterpenes, patchouli alcohol, patchouli, eugenol benzoate, has different specific gravity, namely the higher the concentration of patchouli oil components, the higher the specific gravity [12].

3.3 Refractive

Index The refractive index is the ratio of the speed of light in a substance with the speed of light in air. The refractive index value of patchouli oil is strongly influenced by the density and viscosity of the oil, namely the higher the density of the oil, the higher the refractive index value of the oil [13]. The refractive index of patchouli oil can be determined by the amount of light that passes through a less dense medium to a denser medium so that the light will bend from the normal line [14].

The refractive index value of patchouli oil produced in this test showed satisfactory results in the three samples, in sample A the refractive index value was 1.5111, sample B was 1.5085, and sample C was 1.5086. The average refractive index of patchouli oil is 1.5094 so that all samples of patchouli oil meet the specifications of patchouli oil quality standards. The value of the refractive index shows its constituent chemical components associated with molecular weight. The value of the refractive index following the quality standard is caused by the large content of chemical components with large molecular weights. The more components with large molecular weights such as long-chain components, namely sesquiterpenes or chemical components that contain oxygen, the density of patchouli oil medium will increase so that the incoming light will be difficult to refract. On the other hand, if the refractive index value is not following the quality standard, the molecular weight is low and the density of the medium is also low. This is because the incoming light will be more refracted [15]. The high refractive index value of patchouli oil can also be caused by chemical components, namely oxygenated terpenes in the sample, which are long-chain molecules with unsaturated bonds or contain many oxygen groups [7].

3.4Solubility in Alcohol (Ethanol and Methanol)

According to [11], quality requirements specifythat patchouli oil can be completely dissolved (clear or clear) in alcohol at a ratio of 1:10, meaning that in 1 mL of patchouli oil, 10 ml of alcohol is added or it undergoes mild opalescence (colloidal nature). Patchouli alcohol is one of the groups of oxygenated terpene components contained in patchouli oil characterized by the –OH (alcohol) functional group which has almost the same polarity as the alcohol solvent. Thus, the higher the patchouli alcohol content, the better the solubility of patchouli oil in alcohol and the better the quality of patchouli oil [16].

Testing the solubility of alcohol on patchouli oil samples using ethanol (C_2H_5OH) and methanol (CH_3OH) which is a type of chemical substance that belongs to the alcohol group. The purpose of using ethanol and methanol on the alcohol solubility of patchouli oil is to distinguish between the two solvents whether they have different effects or the same effect. The results of the alcohol solubility test using ethanol and methanol showed that patchouli oil samples tended to be difficult to dissolve in alcohol using either ethanol or methanol. even in the use of ethanol, the solubility of patchouli oil is >1:10, so it requires a larger amount of alcohol to dissolve it. In the use of methanol, the alcohol solubility of sample A and sample B met the SNI quality standards, for sample C it still did not meet. Meanwhile, the use of ethanol showed that the alcohol solubility of the three patchouli oil samples did not meet the SNI quality standards. This is indicated by the number of terpene compounds in the sample. Essential oils containing oxygenated terpene compounds will tend to be more soluble than those containing terpene compounds, because oxygenated terpene compounds are nonpolar compounds that do not have functional groups. So, the higher the terpene content, the more difficult it is to dissolve in alcohol [2].

So it can be concluded that the analysis of the solubility of patchouli oil shows that the use of methanol as a solvent has a different effect when compared to ethanol. It can be seen that ethanol and methanol are polar solvents, although both are polar, methanol is known to be a general solvent and the best when compared to ethanol. Even methanol can bind all chemical analytes found in plants, both polar, semi-polar, and non-polar. Methanol is polar because it has a hydroxyl group (OH) and also can be non-polar because it has an ethyl group (CH_3)as well as non-aqueous methanol as impurities. Meanwhile, ethanol does not have an ethyl group (-CH_3) and contains water as an impurity so that ethanol has greater polarity when compared to methanol and tends to only attract polar chemical analytes [17].

3.5 The Acid Number

principle of acid number is the number of milligrams of potassium hydroxide (KOH) used to neutralize free organic acids in 1 gram of patchouli oil. The specifications for the quality requirements for the acid number of patchouli oil are based on [11], a maximum of 8.00 [14]. The acid number can identify the presence of free acid compounds in the sample. Free acid compounds in patchouli alcohol can affect the quality of patchouli oil and free acid compounds can catalyze oxygen in the outside air so that it will affect the aroma of patchouli oil [18].

The results obtained from testing the acid number of patchouli oil samples were in sample A of 21.0375, sample B of 9.1162, and sample C of 19.6350. This result is not following the specifications for the acid number requirement, even far beyond the maximum limit. The high number of acids can affect the quality of patchouli oil. This is due to the large number of free acid compounds formed due to the chemical components of patchouli oil which are in direct contact with the surrounding air or are in humid conditions, resulting in an oxidation reaction with air (oxygen) catalyzed by light. Therefore, it will trigger the formation of organic acids. This can happen because the distilled patchouli oil is not stored in dark glass bottles or drums with tin-coated inside, but uses plastic jerry cans and is stored freely for an indefinite period waiting for a stable price. Storage time and storage area that is not good allows the interaction between air and oil directly. The interaction of air (oxygen) with oil results in changes in chemical components such as oxygenated terpenes to carboxylic acids through the process of oxidation of alcohol components and hydrolysisof ester components. So that the acid number can be said to be good if the value of the acid number is getting smaller. So it can be concluded that the acid value of the three samples of patchouli oil has poor quality [7].

3.4 The Ester Number

Number The ester number is one of the parameters that indicate patchouli oil has a distinctive aroma, the ester number of patchouli oil states the number of milligrams of potassium hydroxide (KOH) used to saponify the ester in 1 gram of patchouli oil. The ester number in this analysis has different values for each patchouli oil sample, namely in sample A of 28.05, sample B of 21.04, and sample C of 21.04. This result is not following the requirement for an acid number based on the 2006 Indonesian National Standard, which is a maximum of 20.0. The high ester number illustrates that the oil is not easily oxidized so that the distinctive aroma of patchouli oil is more perfect and the aroma lasts longer. However, an ester number that is too high more than the quality requirement of 20.0 indicates the presence of impurities or other substances such as free fat or the presence of additional ingredients mixed in patchouli oil. not good, the ester number is also influenced by the length of distillation, the longer the distillation time the higher the ester number. However, it is seen again with the temperature used. While the patchouli oil samples analyzed did not pay attention to the temperature used. The role of ester number is very important because ester plays a role in determining the aroma of patchouli oil so that the better the ester number, the better the quality of patchouli oil [7].

3.6 Iron (Fe)

Content Iron metal content is one of the causes of dark and cloudy color in patchouli oil, therefore its presence must be known. The principle of determining the content of iron (Fe) in patchouli oil is to compare the absorbance of the sample solution with the standard curve of iron (Fe) solution. Kusuma et al [19], conducted research on the analysis of the heavy metal content of lead (Pb) and mercury (Hg) using atomic absorption spectrophotometry based on the absorption of light by atoms, where the atoms formed are the same as the elements in the cathode lamp. , so the light from the lamp will be absorbed. The rate of absorption depends on the number or concentration of atoms present in the solution.

The content of iron(Fe) in the patchouli oil sample that has been tested and calculated is significantly different, the iron content of sample A is 1.84210 mg/L, for the iron content of sample B must be concentrated 10 times to determine the level of iron (Fe).) which is 0.186842 mg/L, while the highest iron content is Sambel C, which is 6.78950 mg/L. The test results for iron (Fe) content are still following the specifications for acid number requirements based on quality requirements [11], even all test results have levels that tend to be slightly far from the maximum limit. The less iron (Fe) content contained, the better the quality of the oil produced. However, it is not only the iron (Fe) content that causes the color of patchouli oil to darken, the presence of free acid compounds, free fats, and other impurities can also affect it. This is because, if the iron (Fe) content is too high in patchouli oil, it can cause an excessive decomposition of organic matter so that which can affect the quality of patchouli oil.

Journal of Physics: Conference Series **2126** (2021) 012024 doi:10.1088/1742-6596/2126/1/012024

3.7 Patchouli Alcohol and Alpha Copaene Content

quality of patchouli oil that most determines the characteristics of patchouli oil is the chemical nature of patchouli oil, namely the patchouli alcohol content contained in patchouli oil. Patchouli alcohol is a sesquiterpene alcohol chemical compound that can determine the color and distinctive aroma of patchouli oil. The higher the patchouli alcohol content in patchouli oil, the better the quality and the selling price. According to standard specifications [11], the patchouli alcohol content of patchouli oil is at least 30% and the standard quality is based on the Essential Oil Association (EOA), requires 38% [18]. Patchouli alcohol has a white color in the form of crystals, the more patchouli alcohol content will make the color of patchouli oil brighter [15]. The picture below shows the results of the GC-MS test on the three patchouli oil samples.

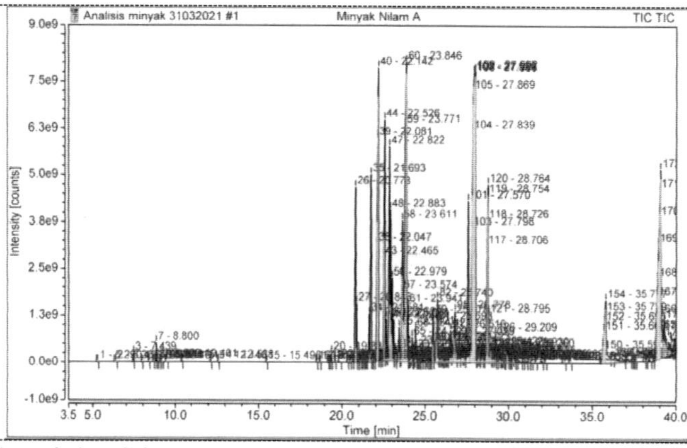

Figure 2. The results of the GC test are the analysis of the separation of compounds (amount of compounds) in sample A of patchouli oil.

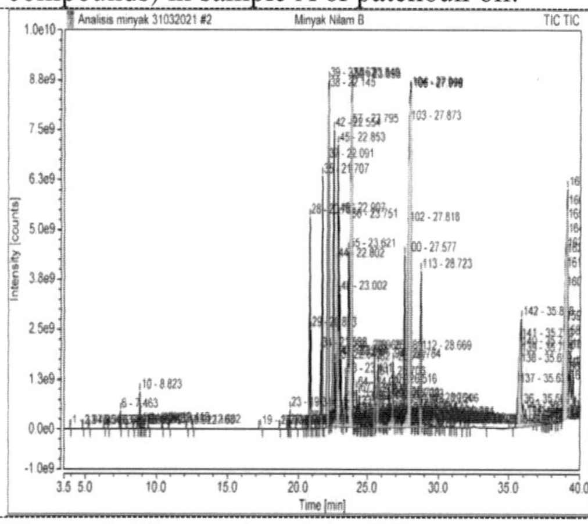

Figure 3. The results of the GC test are the analysis of the separation of compounds (amount of compounds) in sample B oil.

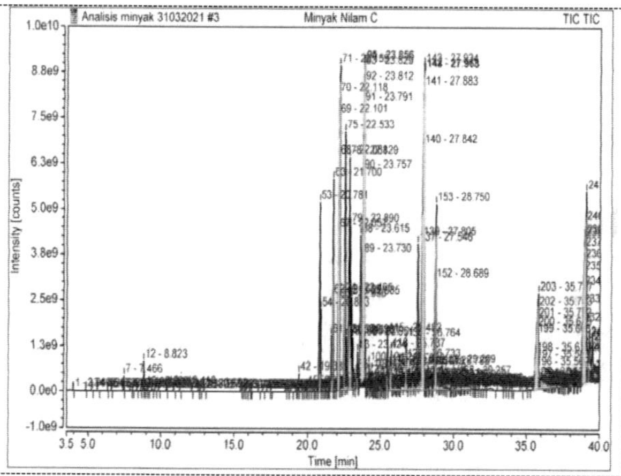

Figure 4. The results of the GC test are analysis of the separation of compounds (amount of compounds) in sample C of patchouli oil.

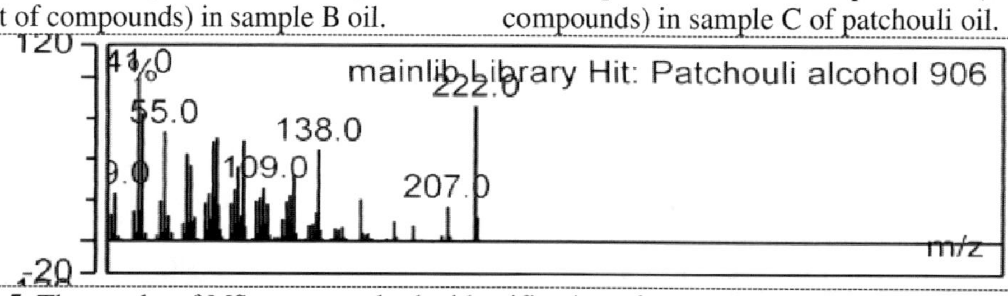

Figure 5. The results of MS test, namely the identification of compounds (relative molecular mass) in sample A of patchouli oil.

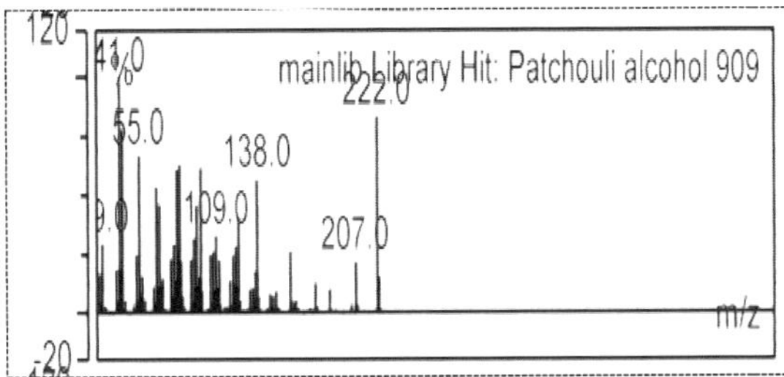

Figure 6. The results of MS test , namely the identification of compounds (relative molecular mass) in sample B of patchouli oil.

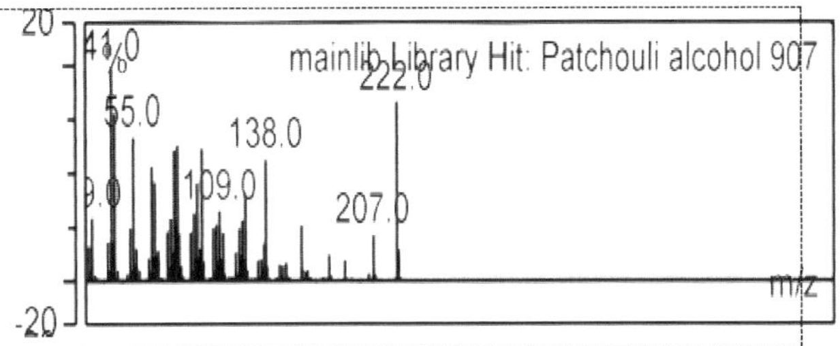

Figure 7. The results of MS test, namely identification of compounds (relative molecular mass) in sample C of patchouli oil.

GC-MS analysis of patchouli oil in sample A showed that there was a patchouli alcohol compound of 16.96%, sample B was 15.52% and sample C is 14.55%, the average patchouli alcohol content among the three samples is 15.67%. So it can be concluded that the patchouli alcohol content of the three patchouli oil samples is below the minimum SNI requirements. The quality of patchouli oil in Tinombala village is still below the standard when compared to the quality of patchouli oil in North Kolaka, the patchouli alcohol content of patchouli oil in North Kolaka ranges from 22%-29%. As for the alpha copaene compound in the patchouli oil sample, it was not detected or did not appear on the chromatogram peaks, so the levels could not be known. The low level of patchouli alcohol is caused by several factors including poor patchouli oil refining and storage techniques, in addition, it is influenced by the condition of the raw material of the patchouli plant being refined such as harvest age and post-harvest handling, the quality of various raw materials and the state of soil nutrients [7].].

Patchouli alcohol content can be analyzed through chromatogram peaks by looking at the relative content of patchouli alcohol which shows the ratio of components in patchouli oil, while patchouli alcohol concentration is expressed by a large peak area. Where the peak area produced by the chromatogram is related to the total concentration of components in the mixture. While the peak area is directly proportional to the concentration. The larger the peak area, the greater the concentration of patchouli oil components [18]. Patchouli alcohol is a water-insoluble compound, but soluble in alcohol, ether, or other organic solvents, it has a boiling point of 280.37oC, and the crystals formed have a melting point of 56°C [20].

The quality and purity of patchouli oil are influenced by many factors, including the type of patchouli plant. In general, the value plant consists of three types, namely Aceh patchouli, Javanese patchouli, and soap patchouli. Judging from the high amount of oil and the composition of the oil produced, the patchouli type of Aceh is superior so that it is more attractive to farmers. However, it is more susceptible to disease when compared to Javanese species [21]. Harvesting of patchouli in Tinombala village ranges from 3-6 months of age by trimming the patchouli plant without leaving branches that can stimulate growth, with harvest time farmers are generally carried out in the morning until noon or do not pay attention to the best time to harvest. Patchouli harvesting is carried out in several stages, the first harvest is done when the patchouli plant is 6 months old and the next harvest is done every 4 months until the plant is three years old. Harvesting should be done in the morning or at night so that the oil content remains high. If the picking is done during the day, it is known that the leaf cells are photosynthesizing so that the rate of oil formation is reduced, the leaves are less elastic and easily torn [22]. In addition, the method of drying patchouli plants can also affect the quality of patchouli oil produced, patchouli plants should be dried in two stages, namely the first stage of patchouli drying in direct sunlight for only 5-6 hours, after that the next stage of patchouli plants is dried in the wind. -air without exposure to sunlight for 3 days. While the drying method used by farmers in Tinombala village is patchouli drying under direct sunlight for 3-4 days until dry and brown plants are obtained. This method can reduce the quality of patchouli plants because some of the oil contained in patchouli plants will easily evaporate or decompose when drying in direct sunlight [7].

In general, patchouli oil distillation is carried out in three ways, namely by boiling (water distillation), this method is considered less efficient and the cost is relatively high. Second, by steaming (water and steam distillation), this method is most widely used by the distillation industry and the last method is steam distillation, this method uses higher pressure and the material is not in direct contact with water as well as the distillation method. which is considered the best when compared to the two previous distillation methods. However, nowadays many patchouli oil refining industries have modified the distillation method to obtain better oil quality [21]. Meanwhile, the distillation method used by farmers in Tinombala village uses water distillation so that there are still many quality parameters of patchouli oil produced that are still below standard. In addition to the method of distillation, the temperature used for refining patchouli oil also affects the oil produced. However, there is no valid information regarding the optimum temperature used for patchouli oil refining. Dalimunthe et al [23] conducted a study on the effect of temperature using direct steam on the quality of patchouli oil with temperature variations of 95oC, 100oC and 105oC obtained the lowest free fatty acid content at 100oC. Lestari et al [24]also researched the effect of the temperature range of fractional distillation to patchouli alcohol content obtained the highest patchouli content in the temperature range of 290-300oC, which is 90.38%. Patchouli alcohol is known to have a boiling point of 150-160 oC at 8 mmHg and 280.73 oC at atmospheric pressure. So it can be said that the temperature used for patchouli oil refining must be ± 100oC because the main compound content of patchouli oil, namely patchouli alcohol, has a high boiling point for evaporation. Mahlinda et al [25] said that the temperature difference can cause the color of patchouli oil to vary. The higher the distillation temperature can cause the color of the oil to turn dark brown. This can be caused by the effect of decomposition or decomposition of hydrocarbon components that occur during the process so that soot appears in the resulting patchouli oil product. Meanwhile, patchouli farmers in Tinombala village did not pay attention to the distillation temperature used. The temperature used is erratic because it uses firewood as a source of fire, the bigger the fire source, the faster the patchouli oil refining will be completed and result in damage to the components that make up patchouli oil, thus affecting its quality and purity.

4. Conclusion

Patchouli oil (*Pogostemon cablin* Benth) produced by Tinombala village, Ongka Malino district, Parigi Moutong district, has color, specific gravity, refractive index, iron (Fe) content which is appropriate so that it meets the quality requirements of the 2006 Indonesian National Standard. Alcohol solubility (ethanol) and methanol), the methanol samples A and B still meet the SNI for sample C not yet complied with. While the alcohol solubility using ethanol (>10:1), the acid number, ester number, and patchouli alcohol content of patchouli oil are still below the quality requirements of the 2006 Indonesian National Standard. Patchouli alcohol content in sample A is 16.96%, sample B is 15, 52%, and sample C is 14.55%.

Acknowledgments

The authors would like to thank all teachers of the Faculty of Teacher Training and Education, Tadulako University.

References

[1] Sukardi, Soeparman S, Argo B . and Irawan Y S 2017 The Effect of Pulsed Electric Field (PEF) on Glandular Trichome and Compounds of Patchouli Oil (Pogostemon cablin, Benth) *J. Eng. Sci. Technol.* **12** 2106–2119

[2] Mukhtar T, Widayat H P and Abubakar Y 2020 Analisis Kualitas Minyak Nilam dan Faktor yang Mempengaruhi Keputusan Petani Dalam Memilih Ketel Penyulingnya *J. Teknol. Dan Ind. Pertan. Indones.* **12** 78–85

[3] Akbar A 2020 Peran Pemerintah Dalam Memaksimalkan Minyak Nilam *Int. J. Gov. Soc. Sci.* **5** 193–202

[4] Hardyanti I ., Septyaningsih D, Nurani I, Agus E and Wibowo P 2016 Analisis Kadar Patchouli Alcohol menggunakan Gas Chromatography pada Pemurnian Minyak Nilam menggunakan Adsorben Zeolit *Semin. Nas. XI Rekayasa Teknol. Ind. Dan Inf. 2016 Sekol. Tinggi Teknol. Nas. Yogyakarta* **8** 392–395

[5] Marpaung, F. P. A. Kesumawati E and Nurhayati 2017 Pertumbuhan Setek Nilam (Pogostemon Cablin Benth) Akibat Pengaruh Dosis Arang Kompos Bioaktif Dan Jenis Zat Pengatur Tumbuh *Seminar Nasional Biotik* pp 347–355

[6] Yahumri, Hidayat P ., Nurjanah S and Mardawati E 2019 Keuntungan dan Kelayakan Usahatani Jagung Hibrida dengan Penerapan Rekomendasi Teknologi Sistem Informasi Tanam Terpadu *AgriHumanis J. Agric. Hum. Resour. Dev. Stud.* **1** 36–42

[7] Rahman A, Rudi L, Ode A . and Wati M . 2019 Analisis Kualitas Minyak Nilam Asal Kolaka Utara Sebagai Upaya Meningkatkan dan Mengembangkan Potensi Tanaman Nilam (Pogostemon sp.) di Sulawesi Tenggara *Akta Kim. Indones.* **4** 133–44

[8] Sekretariat Direktorat Jenderal Perkebunan 2017 *Statistik Perkebunan Indonesia 2016-2018* (Jakarta)

[9] Aisyah Y, Anwar, S H and Annisa Y 2020 Effect of the Fractional Distillation on an Increment Patchouli Alcohol Content in Patchouli Oil *the 2nd International Conference of Essential Oils (ICEO 2019)* pp 80–85

[10] Rahmayanti D, Hadiguna, R A, Santosa and Nazir N 2021 Applying system dynamic for predicting the strengths, weaknesses, opportunities, and treats of Patchouli Oil Agroindustry in West Sumatra *IOP Conf. Ser. Mater. Sci. Eng.* **1041** 1–11

[11] Dewan Standarisasi Nasional 2006 *Minyak Nilam* (Jakarta, Indonesia)

[12] Tutuarima T 2020 Effect Of Natural Fermentation Of By-Products Of Kalamansi Industry On Improvement Of Yield And Quality Essential Oil *AGRITEPA J. Ilmu Dan Teknol. Pertan.* **7** 80–87

[13] Manalu, R A, Patria A and Rohaya S 2019 Peningkatan Mutu Minyak Nilam (Pogostemon cablin) dalam Proses Pemurnian Minyak Nilam Aceh Jaya dan Aceh Selatan dengan Metode Kompleksometri *J. Ilm. Mhs. Pertan.* **4** 310–318

[14] Junaidy R, Fauzia R and Sulaiman I 2019 Peningkatan Kadar Patchouli Alkohol Dalam Minyak Nilam Menggunakan Metode Destilasi Fraksinasi Vakum Bantuan Gelombang Ultrasonik *Maj. Biam* **15** 56–61

[15] Amrullah R, Nurjanah S, Widyasanti A and Muhaemin M 2017 Kajian Pengaruh Rasio Refluks Terhadap Karakteristik Minyak Nilam Hasil Distilasi Fraksinasi *J. Teknotan* **11** 77–88

[16] Nurjanah S, Rosi, D M, Fathoni R P, Zain S, Widyasanti A and Putri I L . 2019 Aktivitas Antibakteri Minyak Nilam (Pogostemon Cablin Benth) Pada Beberapa Tingkat Kadar Patchouli Alcohol *J. Agroindustrial Technol.* **29** 240–246

[17] Ramdani D, Marjuki and Chuzaemi S 2017 Pengaruh perbedaan jenis pelarut dalam proses ekstraksi buah mengkudu (Morinda citrifolia L.) pada pakan terhadap viabilitas protozoa dan produksi gas in-vitro *J. Breed. Sci.* **27** 54–62

[18] Pawestri, A M and Noor F 2019 Effect of Adding Aspergillus niger Mushroom on Patchouli Fermentation Proces *EKSAKTA J. Sci. Data Anal.* **19** 15–25

[19] Kusuma A T, Effendi N, Abidin Z and Awaliah S S 2019 Analisis Kandungan Logam Berat Timbal (Pb) Dan Raksa (Hg) Pada Cat Rambut Yang Beredar Di Kota Makassar Dengan Metode Spektrofotometri Serapan Atom (SSA) *Celeb. Environ. Sci.* **1** 6–12

[20] Sri, H.I, Nurkhalisa S and Maghfiroh H 2017 Comparison of bentonite and zeolite as adsorbent purification process of patchouli oil (Pogostemon cablin) *J. Sci. Innov. Res.* **6** 87–90

[21] Sahwalita and Herdiana N 2016 *Panduan Budidaya Nilam (Pogostemon cablin Benth.) dan Produksi Minyak Atsiri* (Sumatera Selatan)

[22] Supendy R 2016 Analisis Ekonomi Dan Manajemen Perbaikan Sistem Usahatani Nilam Di Desa Kabupaten Bombana Sulawesi Tenggara (Kasus Di Desa Babamolingku) *J. Agrol.* **23** 164–173

[23] Dalimunthe H, Harahap L . and Munir A P 2015 Effect of Destillation Temperature on Quality and Yield of Patchouli Oil of Direct Steam Destillation Type *J. Rekayasa Pangan Dan Pertan.* **3** 377–381

[24] Lestari P, Nurjanah S and Mardawati E 2020 Pengaruh Rentang Suhu Distilasi Fraksinasi Terhadap Kadar Patchouli Alcohol (PA) Pada Minyak Nilam *AgriHumanis J. Agric. Hum. Resour. Dev. Stud.* **1** 36–42

[25] Mahlinda, Arifiansyah V and Supardan M . 2019 Modifikasi Alat Penyuling Uap untuk Peningkatan Rendemen dan Mutu Minyak Nilam (Pogostemon cablin Benth) *J. Chem. Eng. Enviromental* **14** 28–35

Development of a microcontroller-based instrument for measuring liquid density

F Handayani[1*], U Wahyono[1], and S Saehana[1]

[1]Sub-Department of Physics Education, Faculty of Teacher Training and Education, Universitas Tadulako, Jl. Soekarno Hatta KM. 9 Kampus Bumi Tadulako Tondo Palu-Central Sulawesi, Indonesia.

[*]Corresponding author: handayanifinda@gmail.com

Abstract. The purpose of this research was to develop a practical tool to measure the density of liquids. This research was a research and development research that referred to the development model of Borg and Gall. The research instrument used was a questionnaire with a 4-Likert scale. Evaluation phase included the assessment of media validator, students' responses, and laboratory test. The product quality data obtained were analyzed using descriptive analysis. The quality of the product based on the analysis of the media expert's assessment obtained a value of 3.30 which was in very good category. Meanwhile, the results of student responses in the limited-testing obtained a value of 3.30 which showed that they strongly agree with the product's usability. The test results and questionnaire analysis indicated that the microcontroller-based density determination practicum tool developed was feasible to use. From the experimental results, the density value of water was 0.95 gr/cm^3 with a standard deviation of ±0.044 and a relative uncertainty value of 0.22%. The density value for salt water was 1.04 gr/cm^3 with a standard deviation of ±0.011, compared with the literature which is 1.10 gr/cm^3. Moreover, the density of coconut oil measured was 0.76 gr/cm^3 with a standard deviation of ±0.045, while the literature is 0.84.

1. Introduction

Practicum is required to help students understand the real application of the concepts of physics. It provides opportunities for students to find theories and prove theories. Through practicum, students can see the demonstration of the physics concepts and principles. In addition, hands-on activities can increase students' creativity and skills and subsequently improve their learning performance [1], motivation [2], engagement [3], and curiosity [4,5]. These are the attribute to intrigue self-studying. The involvement of hands-on activities also increases students' motivation to study technology and engineering related areas for their higher education [6].

Topic about density has been learned by students since junior high school. This quantity is used in many physics' concepts. However, in most cases, the value of density has been provided during problem solving because the focus of problem is the other quantities.

According to its definition and formula, determining the density of a substance can be done by measuring the mass per unit volume. For liquid, there are some methods that can be used to determine the density without calculating the mass and volume of the liquid. It uses the principles of Archimedes with a simple mathematical ratio [7,8]. It was based on a study of Hughes [9] that explored a more accurate and faster method to measure small object volume. In addition, Putranta et al [10] designed a simple measuring tool for density based on Hook's Law using simple materials, such as pipe, rulers, springs, and wooden table. However, the experiments related to density are commonly conducted manually, especially for science learning in schools by measuring the mass and calculating the volume of an object. Although this activity can improve their scientific skills, it takes more time.

Since the advancement of technology and science, they get strongly intercorrelated. Most of the activities in education sector, including laboratory activities have widely involved the utilization of technology. In higher education, most of measuring instruments are digital. However, there has not been a digital measurer of density in the researchers' department. Some researchers have been conducted to develop a laboratory tool using the technology-based. For example, Warsito [11] conducted research to measure density using a sample of a liquid with the help of an LDR sensor. For the measurement process, a Visual Basic 6.0 program was used which was displayed on a monitor. The results obtained were close to the literature. Moreover, Pande [12] conducted a study on the realization of a liquid density measuring instrument based on the hydrostatic pressure method using a photodiode sensor based on an Atmega 8535 microcontroller. The result indicated that the accuracy value was 97%. Megantoro et al [13] designed a digital liquid density meter based on Arduino, while Roy et al [14] used iezoelectric Micromachined Ultrasound Transducers (PMUTs). All the results of the previous research showed that the density measurement were accurate.

Therefore, we intended to develop a practical tool for determining the density of liquids based on a microcontroller. The components required for measuring instruments based on microcontroller are relatively low cost and low power system [15]. Moreover, our intention was that this product can provide more accurate results within a short time. By saving their time would give the students more time for discussion and brainstorming of their gained results.

2. Methods

This research was a Research and Development (R&D). The research model adapted the model of Borg and Gall [16]. The steps of the research model are presented in Figure 1.

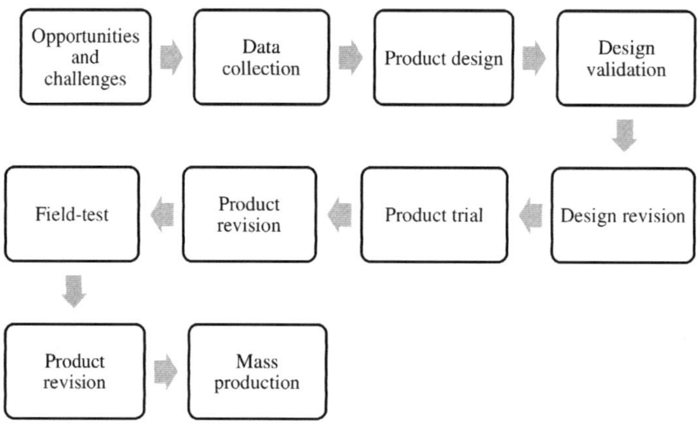

Figure 1. Stages of research and development (R&D) methods according to Borg and Gall

However, this study only conducted six stages, including the Potential and Problems, Data Collection, Product Design, Product Validation, product revision and product trial (limited test). The subjects of this study were students of the Sub-Department of Physics Education, Faculty of Teacher Training and Education, Tadulako University. The data were collected through questionnaire distribution to a media expert, and 10 of undergraduate students as prospective users for the limited-test. The questionnaire consists of questions that can be used to assess the quality and feasibility of the resulting learning media. Students were given the questionnaires after using the developed product in the basic physics lesson. Before testing to students, the product has been undergone field-test in laboratory to see the density value of the measured substance. The data were analyzed using descriptive analysis with the maximum average value of 4.0. It was based on the 4-Likert scale of the questionnaire [17].

3. Results and Discussion

3.1. Initial Stage

The initial stages of this research involved designing the product, gathering the components, and arranging the practical tools according to the initial design. Manufacturing stage was started with designing the mechanical parts of the tool to make all sensors function properly. For example, the researchers had to consider on how the shape of the plywood to make the load cell and the layout the sensors and other components to make them function properly. It was a critical stage since the mechanical shape of the components greatly affects the accuracy of the measuring instrument. The next stage was managing the electrical components. At this stage, it was also significant to carefully scrutinize the load cell and ultrasonic sensor, especially on how the sensor were connected with the battery with the driver and with the microcontroller used (Arduino). They will serve as a volume meter. Moreover, after the practicum tool has been arranged properly, the next stage was programming. The programming language used was the C++ language in the Arduino application. The developed product was displayed as Figure 2.

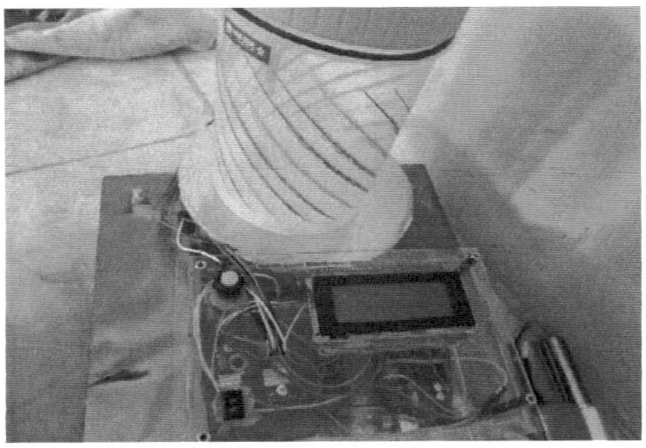

Figure 2. Display of practical tools

3.2. Lab Manual

In this study, we also provided a lab manual to help students operating the practical tool and to allow self-study between them. Figure 3 is an example of the descriptions of materials used during the experiment. Figure 3 provides tools and materials used including a heavy sensor, Arduino uno, rainbow cable, battery, LCD, i2c, and plastic jars.

3.3. Laboratory Test Result

Laboratory test was done after the product was considered ready to be used. Laboratory tests were conducted to three kinds of liquid. They were water, coconut oil, and salt water. The detailed comparison between the density value based on the product measurement and the literature were displayed in Table 1. From the experimental results, the density value of water is 0.95 gr/cm^3 with a standard deviation of ±0.044 and a relative uncertainty value of 0.22%. The density value for brine obtained is 1.04 gr/cm^3 with a standard deviation of ±0.011 while the density of coconut oil in the literature is 1.10 gr/cm^3. For coconut oil, the density obtained is 0.76 gr/cm^3 with a standard deviation of ±0.045 and the density according to the literature is 0.84. Based on the laboratory test, there were some revisions made. First, the quantities displayed on the LCD previously were mass, volume, height and radius. It was changed to mass, volume, height and density.

Table 1. The density value of the research objects to the literature

Liquid	Density (ρ) gr/cm^3 (literature)	Density (ρ) gr/cm^3 (measurement)	Standard deviation (σ)
Water	1.00	0.95	0.044
Coconut oil	0.84	0.76	0.045
Salt water	110	1.04	0.011

LAB MANUAL DEVELOPMENT OF MICROCONTROLLER-BASED MICROCONTROLLER-BASED PRACTICUM TOOLS

A. Purposes
   1. Students are able to understand the concept of density.
   2. Students can analyze the data from the density measurement of a liquid

B. Components

1.	Heavy sensor	
2.	Arduino Uno	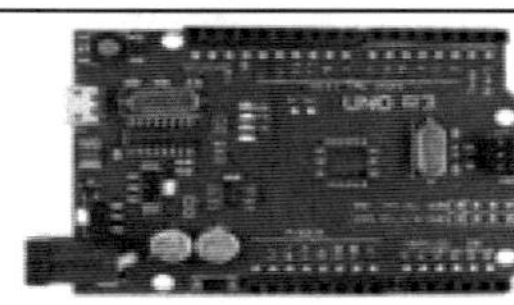
3.	Rainbow cable	
4.	12 volt battery	
5.	LCD	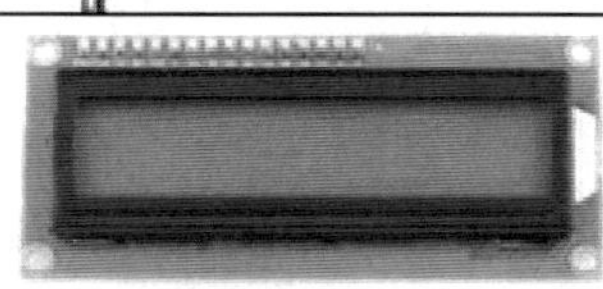
6.	i2c	
7.	Plastic jar	

Figure 3. Display of the Lab Manual for practicum tools and materials used

3.4. Expert Evaluation

The media assessment was used to evaluate the feasibility of the practicum tools and the feasibility of the lab manual. Some suggestions that were given by the expert included the improvement in product quality by stabilizing the numbers that appear on the LCD. The expert reckoned that the numbers kept changing. It was an imperative comment since the numbers indicated the measurement results. Further, the expert suggested to apply the product to other kinds of liquid to ensure the measurement accuracy of the product. In addition, the lab manual was required to cover conclusions and suggestions columns.

The results of expert assessment showed that the average score of product feasibility was 3.50 which was in very good category. In the ergonomic aspect, the average value was 3.00 with good category. In terms of the suitability of the device with the concept of physics, the average value was 3.20 in good category. Meanwhile, it got a perfect score of 4.0 for the aspect of time efficiency. Meanwhile, the results of the manual lab assessment in the aspects of the content presentation and language obtained 2.50 and 3.50 respectively. These results denote that the tool was feasible to be tested in the field with revisions.

3.5. Limited-test (Field-test)
The limited test was conducted to measure the aspects of the physical appearance of the device, ergonomic, learning motivation, and operational and performance. The physical appearance gained an average score of 3.00 and was categorized as agree. Meanwhile, the rest of the aspects, the ergonomic, learning motivation, and operational and performance aspects were in strongly agree category with the average of 3.55, 3.37 and 3.45, respectively. Overall, students strongly agree with the feasibility and usability of the developed product, with the average score of 3.31. these results were in accordance with the assessment result of the manual lab. The lab manual was assessed based on the content presented and language and both aspects obtained the average score above 3.0.

4. Conclusion
A practical tool for determining the density of liquids based on a microcontroller has been made. The results of the assessment from the media expert and undergraduate students indicated that the tool was feasible to use. Besides the positive feedback and respondents from the respondents, the feasibility of the product also was shown from the laboratory test. Although the density values of the measurement were not exactly the same as the literature, the small value of standard deviation should be considered as the feasibility of its application.

Acknowledgment
We kindly thank Hibah Penelitian Terapan Kompetitif Nasional DIKTI and Program Kompetisi Kampus Merdeka (PK-KM) of Ministry of Education, Culture, Research and Technology of The Republic of Indonesia, under the funding 302/E4.1/AK.04/PT/2021 and 119/E1/KM.05.03/2021, respectively.

References
[1] Yehya F M, Barbar A M and Abou-Rjelil S 2019 Learning with simulations: Influence of a computer simulation with hand- on activities on students' learning of the physics capacitors' concepts *Res. Soc. Sci. Technol.* **4**

[2] Chen J C 2020 Developing a cycle-mode POED model and using scientific inquiry for a practice activity to improve students' learning motivation, learning performance, and hands-on ability *Interact. Learn. Environ.*

[3] Morris B J, Owens W, Ellenbogen K, Erduran S and Dunlosky J 2019 Measuring informal STEM learning supports across contexts and time *Int. J. STEM Educ.* **6**

[4] Waris A 2015 *Pengembangan Alat Praktikum Sederhana Konsep Listrik Magnet untuk Siswa SMP Di Daerah Terpencil* (Universitas Tadulako)

[5] Kibga E S, Gakuba E and Sentongo J 2021 Developing Students' Curiosity Through Chemistry Hands-on Activities: A Case of Selected Community Secondary Schools in Dar es Salaam, Tanzania *Eurasia J. Math. Sci. Technol. Educ.* **17**

[6] Filho E F S, Pena F L A, De Albuquerque M C S, Farias P C M A and Reis M 2016 Hands-on activities in experimental high-energy physics for attraction and motivation to engineering careers *Int. J. Electr. Eng. Educ.* **53** 72–86

[7] Hughes S W 2006 Measuring liquid density using Archimedes' principle *Phys. Educ.* **41**

[8] Kiriktaş H, Şahin M, Eslek S and Kiriktaş I 2018 A new approach to determine the density of liquids and solids without measuring mass and volume: Introducing the solidensimeter *Phys. Educ.* **53**

[9] Hughes S W 2005 Archimedes revisited: A faster, better, cheaper method of accurately measuring the volume of small objects *Phys. Educ.* **40** 468–74

[10] Putranta H, Wiyatmo Y, Supahar X X and Dwandaru W S B 2020 A simple liquid density measuring instrument based on Hooke's law and hydrostatic pressure *Phys. Educ.* **55**

[11] Warsito, Pauzi G A and Jannah M 2013 M.Jannah, Warsito, dan G.A. Pauzi. Analisis Pengaruh Massa Jenis Terhadap Kualitas Minyak Goreng Kelapa Sawit Menggunakan Alat Ukur Massa Jenis dan Akuisisinya Pada Komputer. *Semirata FMIPA Universitas Lampung*

[12] Pande Av 2017 *Realisasi Alat Ukur Massa Jenis Zat Cair Berdasarkan Metode Tekanan Hidrostatik Dengan Menggunakan Sensor Fotodioda Bebasis Mikrokontroler ATmega 8535* (Universitas Lampung)

[13] Megantoro P, Widjanarko A, Rahim R, Kunal K and Arfianto A Z 2020 The design of digital liquid density meter based on arduino *J. Robot. Control* **1** 1–6

[14] Roy K, Gupta H, Shastri V, Dangi A and Pratap R 2018 Fluid Density Sensing Using PMUTs *Proceedings of IEEE Sensors* vol 2018-October (Institute of Electrical and Electronics Engineers Inc.)

[15] Somantri Y, Wahyudin D and Fushilat I 2018 Personal Computer-less (PC-less) Microcontroller Training Kit *IOP Conference Series: Materials Science and Engineering* vol 306 (Institute of Physics Publishing)

[16] Sugiyono 2012 *Metode Penelitian Pendidikan: Pendekatan Kuantitatif, Kualitatif, dan R&D.* (Bandung: Alfabeta)

[17] Arikunto S 2011 *Prosedur Penelitian Pendekatan Suatu Praktik* (Jakarta, Indonesia: Rineka Cipta)

The use of water spinach plants (Ipomoea aquatica Forsk.) for phytoremediation of hospital waste

Suherman[1], S Rahmawati[1*], I Said[1], Nurbaya[1], S Armiyanti[1], and N Thamrin[2]

[1]Chemistry Education, Department of Mathematics and Natural Science Education, Faculty of Teacher Training and Education, Tadulako University, Palu 94118 – Indonesia
[2]English education study program, Faculty of Teacher Training and Education, Tadulako University, Palu 94118 – Indonesia

[*]Corresponding author: suherman@untad.ac.id

Abstract. Water spinach (Ipomoea aquatica Forsk.) has the ability to accumulate pollutants so that it can be useful for wastewater treatment processes. The purpose of this study was to determine the phytoremediation technique in hospital wastewater using kale in improving the quality of hospital wastewater and to determine the level of accumulation of Pb and Cr metals in plants. The study used a completely randomized design with four treatments, namely treatment 0 days (without treatment), 2 days, 4 days and 6 days. Parameters include the physical and chemical quality of the waste as well as the content of Pb and Cr in the waste and plants. To determine the effect of phytoremediation on the quality of household waste and metal content, ANOVA analysis and further testing with the BNT test level of 5% were used. Phytoremediation with kale was able to reduce the temperature of the waste on the 4th day of treatment and dissolved oxygen on the 4th treatment and increase the pH. The accumulation of Pb and Cr on each day is different. The highest accumulation of Pb and Cr occurred on day 6, namely 0.1587 mg/L for Pb metal and 0.2167 mg/L for Cr metal. Kale plants are very possible to be used in the phytoremediation process.

1. Introduction

Environmental pollution problems related to water pollution are experienced by many developing countries such as Indonesia. This pollution needs to be anticipated to overcome the level of pollution. Please note that the source of water pollution comes from urban waste and industrial waste. One of the urban wastes comes from hospital wastewater [1].

The hospital is one of the efforts to improve health which consists of a medical center and a doctor's practice which is also supported by other units, such as operating rooms, laboratories, pharmacy, administration, kitchens, laundry, waste and waste management, as well as providing education and training [2].

Hospital is a complex health facility. One of them is the Anutapura Palu General Hospital which has Type B, which has a fairly solid activity in carrying out complex services to patients [3].

Hospital waste can be divided into two, namely liquid waste and solid waste. Liquid waste can contain organic and inorganic materials which are generally measured and parameters BOD, COD and TSS, while solid waste consists of easily decomposed waste, infectious waste, and others. These wastes are likely to contain pathogenic microorganisms or hazardous and radioactive toxic chemicals that can cause infectious diseases which can then be spread to the hospital environment [4].

Treatment of waste generated from hospitals is necessary to minimize its impact when released into the environment. Waste treatment aims to reduce the toxic materials contained in the waste to a minimum or even zero (0)% before the waste is disposed of. One of the efforts that can be made for the hospital waste treatment process is the use of natural resources, namely utilizing the types of aquatic plants that grow in the sewers around the settlements [5].

The use of aquatic plants in the wastewater treatment process causes the process of ion exchange and absorption [6]. One of the treatment methods for tackling water pollution by using these aquatic plants as a medium for absorbing waste is the phytoremediation method [7]. One of the aquatic plants that can be used as phytoremediation agents is water spinach. This plant has the ability to absorb pollutants and heavy metals dissolved in the growing media so that its content decreases or the quality of wastewater increases [8].

The use of kale for phytoremediation has been carried out on various wastes, including: tapioca liquid waste, batik waste and tofu industrial waste [8]. Therefore, in this study, phytoremediation was carried out on Anutapura hospital waste. The purpose of this study was to determine the phytoremediation technique in hospital wastewater using water spinach (*Ipomoea aquatica* Forsk) in improving the quality of hospital wastewater and to determine the level of accumulation of Pb and Cr metal levels in water spinach plants (*Ipomoea aquatica* Forsk).

2. Methods

2.1 Pre-research (Preparation Stage)

2.1.1. Preparation stage for water spinach plants (Ipomoea aquatica Forsk.)

The water spinach plant (Ipomoea aquatica Forsk.) which were used in this study was taken from rice fields in the biromaru area and then carried out the acclimatization process on water growing media. Acclimatization is a stage of adaptation of plant tissue culture results to the new environment. Acclimatization can be referred to as the stage of adjustment, before the plant is finally able to live in the field [9]. This acclimatization process aims for plants to adapt or adapt to new environments or unusual conditions, such as changes in temperature, humidity and light intensity. The procedure for acclimatization of kale is to prepare ingredients in the form of fresh water spinach and water, and prepare tools in the form of a bucket, steroform as a support for kale, and a measuring cup. Fill the bucket with 2 liters of water. Choose kale with the same size and arrange in a bucket using steroform as a support. Allowing water spinach plants to grow for 7 days or 1 week [10].

2.1.2. Hospital liquid waste preparation phase the hospital

Liquid waste used in this study was taken at one of the hospitals in Palu City, namely Anutapura General Hospital. Liquid waste is taken from several sources of hospital disposal using containers in the form of jerry cans and then taken to the research location, namely the chemistry laboratory, faculty of teacher training and education, then mixed thoroughly and analyzed for its physical and chemical properties. The next step is the liquid waste is filled into 12 glass containers according to the treatment, as much as 1 liter for each container.

2.2 Research Procedures

2.2.1. Phytoremediation process

Water spinach plants that have been acclimatized and then grown are transferred into a glass container that already contains 1 L of hospital liquid waste. Where plant maintenance in liquid waste media is carried out according to a completely randomized design with the treatment that the plants are maintained in liquid media for 2x24 hours for three repetitions, for 4x24 hours for three repetitions, for 6x24 hours for three repetitions and control treatment. After that, the liquid waste samples were re-analyzed to determine the physical and chemical quality. The physical quality measured include temperature, conductivity and turbidity, while the chemical quality include pH and O_2 dissolved (DO).

2.2.2. Analysis of Pb and Cr content in plant samples was carried out by atomic absorption spectrophotometry method.

Preparation of plant samples to be analyzed was carried out by crushing 1 g of plant parts (roots, stems and leaves). Added 10 mL of HNO_3. Heated at 90 °C for 20 minutes. The temperature is raised to 150 °C for 1 hour. Added distilled water until the volume reaches 25 mL. The levels of lead (Pb) and chromium (Cr) were tested using an atomic absorption spectrophotometric (AAS) instrument. The data obtained were analyzed statistically using Anova and further tested [6].

3. Result and Discussion

This research is to carry out a phytoremediation technique in Anutapura Palu hospital wastewater by utilizing a plant, namely water spinach (*Ipomoea aquatica* Forsk) to improve the quality of hospital wastewater and accumulate Pb and Cr metal levels in the waste. Phytoremediation is the treatment of waste by using plants and their parts which function to reduce pollutants in the environment contaminated with waste. This phytoremediation process is one way to reduce the concentration of pollutants in the waters. The phytoremediation method can reduce and neutralize Pb heavy metal content in soil and water [11].

Water spinach that has been taken from the initial location is then acclimatized which aims to adapt the plant to the surrounding environmental conditions, namely temperature, climate, temperature and others before the plant is finally able to live in the actual environment at the phytoremediation test research stage. This acclimatization was carried out for ± 7 days by moving water spinach from the initial location to the study site, then the plants were cleaned first of dirt attached to the roots, stems and leaves. After that, the plants are put in a container that has been filled with water [12].

Water spinach plants that have been acclimatized are then grown by being transferred into a glass container that already contains 1 L of hospital liquid waste. Where plant maintenance in liquid waste media is carried out according to a completely randomized design with the treatment that the plants are maintained in liquid media for 2x24 hours for three repetitions, for 4x24 hours for three repetitions, for 6x24 hours for three repetitions and control treatment. After that, the liquid waste samples were analyzed to determine the physical and chemical quality. The physical quality measured include temperature, conductivity and turbidity, while the chemical quality include pH andO_2 dissolved (DO).

Table 1. Results of measuring the physical and chemical quality of hospital effluent

	Physical Quality					Quality Chemical	
Treatment	(°C)	(NTU)	(μS)	Color	Smell	pH	DO (mg/L)
Day 0	30	2.19	2, 32	White Cloudy	Smell	7.91	2.07
Day2	31	3.91	2.94	White Cloudy	Smell	7.84	2.00
Day 4	25	4.38	4.51	White Cloudy	Less Smelly	8.33	1, 29
Day 6	30	5.63	6.31	White Cloudy	Less Smelly	8,21	3,66

In general, utilization of water spinach plants can alter the physical and chemical wastewater hospitals (Table 1). Based on the data in the table, the temperature measurement varies from 25°C to 31°C. The increase in temperature can increase the rate of ion diffusion to the roots of kale plants including Hg ions and Cr ions [13]. From the data in Table 1 it can also be seen that the turbidity value increased from the first day to the sixth day (2.19 – 5.63 NTU). This shows that the water spinach plant has not been able to reduce the level of water turbidity but this condition is still in good water stability and far below the threshold range of a waters where the maximum threshold for turbidity of aquatic biota is 30 NTU so that in this condition it is still optimum for life [14].

The results also showed an increase in the conductivity value, which ranged from 2.32 to 6.31 S/cm (Table 1). This value is still very good and meets the standard requirements of <700 that have been set by FAO (Food and Agriculture Organization) for agriculture. The data in Table 1 shows that the pH of the liquid waste also increased at the end of the observation. It can be concluded that kale has the potential to increase the pH of hospital wastewater. The pH value ranges from 7.91 to 8.34. The pH value obtained is still in the normal pH range for biota life in a waters, which is 6-9 [15,16]. The pH parameter needs to be studied during phytoremediation because it can affect the solubility of nutrients that cause plant growth. The increase in pH may be due to the oxidation of organic matter during the phytoremediation process.

Dissolved oxygen is a basic requirement for plant life in water. The ability of water to maintain the concentration of O_2 minimum needed in life greatly affect plant life in the water. The concentration ofO_2 dissolved will be at least microbial life can not be less than 6 ppm. Temperature and atmospheric pressure are factors that can affect various concentration O_2 in the state of saturation. The maximum of dissolved at 1 atm to a temperature of 24-28 °C range between 8.5 to 9.9 ppm. To theOconcentration$_2$ dissolved is too low and too high can interfere in aquatic biota.

Based on research data, it shows that kale has a high adaptability to physical and chemical waste factors, with the fact that in the field it is able to grow in a very low dissolved oxygen range (1.29-3.66 ppm) below normal (8.5-9.0 ppm).

First of all, before testing the phytoremediation of waste by plants, a preliminary test was carried out to determine the content of pollutants such as dangerous heavy metals contained in the waste using the atomic absorption spectrophotometry (AAS) method at Chemistry Laboratories, Faculty of Mathematics and Natural Sciences. The test results obtained that there is a level of waste Pb of 5.829 ppm and waste Cr of 5,000 ppm.

Changes in the physical condition of water spinach plants during the phytoremediation process were clearly visible in each treatment. Starting from the condition of the leaves that are increasingly yellowing in color and withering over time of contact, accompanied by a decrease in the volume of wastewater. In general, the leaves begin to turn yellow due to exposure to heavy metals, this indicates the inhibition of chlorophyll formation in plants. The presence of metals (particularly Pb and Cr) contributes to the disruption of the photosynthetic process in this case disrupting the function of enzymes that act as chlorophyll biosynthesis.

Phytoremediation test research on the absorption of heavy metals lead (Pb) and chromium (Cr) in water spinach plants (*Ipomoea aquatica* Forsk.) was conducted to determine the accumulation ability of these plants. The study was conducted for 6 days. The test for the absorption of lead (Pb) and chromium (Cr) in plants was carried out four times, namely on the 0, 2, 4 and 6 days. The absorption rate of heavy metals lead (Pb) and chromium (Cr) is known from the time of contact or exposure of plants to heavy metals during the study.

The ability of absorption of heavy metals Pb and Cr by water spinach plants which were harvested on days 2, 4 and 6 were different. The longer the time to harvest the kale, the greater the levels of heavy metals Pb and Cr absorbed by the plant. On the second day of phytoremediation of Pb metal, the absorbed metal content decreased by 0.0476 mg/L, on the 4th day it decreased by 0.111 mg/L, and on the 6th day it decreased by 0.1587 mg/L (Fig 1). Microorganisms that help reduce concentration are around the roots of water spinach. Pb that is absorbed by the roots accumulates in it, the rest is carried to the stomata. In the water spinach body there is a metabolic process of decomposition of Pb mixed with oxygen to produce PbO_2 which, if released into the air, does not experience environmental pollution.

Phytoremediation of Cr metal, found a decrease in levels on day 2 of 0.0500 mg/L, day 4 decreased by 0.1500 mg/L and on day 6 decreased by 0.2167 mg/L (Fig 2). The process of decreasing by rhizofiltration and phytoextraction. Cr which is absorbed by water spinach through the roots is carried to the leaves to be released into the air through the stomata. The decrease in the concentration of Pb and Cr is influenced by the activity of microorganisms that play a role in helping to reduce the concentration in large quantities. Water spinach is a plant that can accumulate Pb and Cr in high concentrations so that it can be used as a phytoremediation plant for polluted substances. Effect of the residence time is directly proportional to decreasing the concentration of Pb and Cr (see Figure 1 and Figure 2) and the large number of water spinach greatly affect the absorption of the concentration of Pb and Cr.

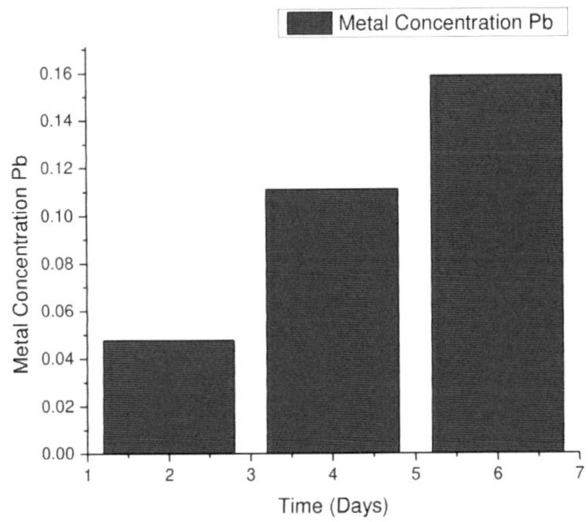

Figure 1. Graph of contact time relationship with metal concentration Pb

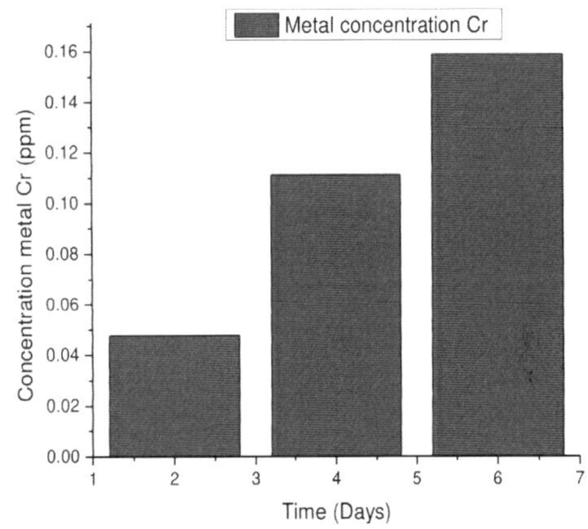

Figure 2. Graph of the relationship contact time with concentration of Cr

Several factors that affect water spinach in heavy metal absorption are metal accumulation in plants not only depending on the heavy metal content in water and soil, but also soil chemical elements, metal types, soil pH, and plant species. In relation to harvesting time, the longer it is harvested, the more heavy metals will accumulate in the body of the water spinach plant [17].

Table 2. ANOVA test results for reducing Pb and Cd metal levels with water spinach (*Ipomoea aquatica* F.)

	Sum of Squares	df	Mean Square	F	Sig
Pb (ppm) Between Groups	0.019	2	0.009	18.494	0.003
Within Groups	0.003	6	0.001		
Total	0.022	8			
Cr(ppm) Between Groups	0.042	2	0.021	76.000	0.000
Within Groups	0.002	6	0.000		
Total	0.044	8			

Based on Table 2 shows the results of data calculations using Analysis of Variance (ANOVA)

are the Fcount = 18,494 and 76,000 when compared to the distribution list table F. The Fcount value is greater than Ftable with a significant level of = 0.05 (5%) in degrees independent (df) 2 and 6 so that Ftable = 5.14. Thus Fcount > Ftable or p (0.000) < 0.05 in the treatment, meaning that H_0 is rejected. This shows that there is an effect of water spinach plants (*Ipomoea aquatica* F.) on the reduction of Pb and Cr metal levels in hospital waste, or various weights of water spinach (*Ipomoea aquatica* F.) have a significant effect on reducing Pb and Cr metal levels. From the coefficient of diversity (KK) obtained a value of 15.00822%. The resulting KK value is less than 20%, then it is continued by using the BNT test (Least Significant Difference). The results of the calculation using the test BNT (Least Significant Difference) is 0.035901. This value will be compared with the average that exists in the measurement results of reducing the levels of Pb and Cr metals. If the average result is close to the BNT value, there is no significant difference, but if the average has a large difference with the BNT value, there is a significant difference in the test.

The LSM/BNT test table shows that based on the Least Significant Difference (BNT) test, there are nine treatments that are in different subsets, meaning that the value is less than 0.03. The effect of the phytoremediation method using water spinach (*Ipomoea aquatica* F.) with Ptreatments$_1$, P_2, P_3 and the control there was a significant difference in the effect of reducing Pb and Cr metal levels in hospital wastewater.

4. Conclusion

Water spinach (*Ipomoea aquatica* Forsk.) is a plant that has the ability to accumulate Pb and Cr metals in high concentrations in hospital waste samples so that it can be used as a phytoremediation plant for polluted substances. The metal content of Pb and Cr was different in each treatment. The highest metal content was found at the longest harvest time, namely day 6 of 0.1587 mg/L for Pb metal and 0.2167 mg/L for Cr metal. Cr accumulation was higher than Pb in all treatments.

Acknowledgement

The author would like to thank the Head of the Chemistry Laboratory of FKIP Universitas Tadulako and the Chemistry Laboratory of FMIPA Untad who have facilitated this research.

References

[1] Ferronato N and Torretta V 2019 Waste mismanagement in developing countries: A review of global issues *Int. J. Environ. Res. Public Health* **16**

[2] Ramadhani C, Gafur A and Idris F P 2019 Efficiency of wastewater treatment at pelamonia hospital in Makassar and syekh yusuf hospital in Gowa *Celeb. Environ. Sci. J.* **1** 27–35

[3] ousefli Z, Nasiri F and Moselhi O 2017 Healthcare facilities maintenance management: a literature review *J. Facil. Manag.* **15**

[4] Said N I and Wahjono H D 1999 *Hospital Wastewater Treatment Technology with "Aerobic-Aerobic Biofilter" System* (Jakarta, Indonesia)

[5] Adisasmito W 2014 *Hospital environmental management system* (Jakarta, Indonesia)

[6] Lestari W 2013 *Use of Ipomoea aquatica Forsk for phytoremediation of household waste* (Riau University)

[7] Dewi F, Faisal M and Mariana 2015 The efficiency of phosphate absorption in laundry waste using water spinach (Ipomoea aquatic forsk) and jeringau (Acorus calamus) USU *J. Chem. Eng.* **4** 7–10

[8] Sinulingga N N K and K A E B 2015 Phytoremediation of metallic mercury (Hg) in aqueous media by water spinach (Ipomoea aquatica Forsk.) *Environ. Biol.* **2** 75–81

[9] Nurmalinda, Yuliansyah A T and Prasetya A 2018 Aklimatisasi Tanaman Lemna Minor Dan Azolla Microphylla Terhadap Lindi Tpa Piyungan Pada Tahap Awal Fitoremediasi *Pros. Pertem. dan Present. Ilm. Penelit. Dasar Ilmu Pengetah. dan Teknol. Nukl.* **ISSN 0216-**

[10] Handayani A D, Herman T, Fatimah S, Setyowidodo I and Katminingsih Y 2018 Inquiry based learning: A student centered learning to develop mathematical habits of mind *Journal of Physics: Conference Series* vol 1013

[11] Tarigan-Sibero N H B, Wijayanti N P P and Perwira I Y 2019 Phytoremediation of heavy metal lead (Pb) by kiapu plant (Pistia stratiotes) based on mass balance analysis Current *Trends Aquat. Sci.* **2** 89–95
[12] Ghiovani D and Tangahu B O 2017 Phytoremediation of water contaminated by laundry waste using apu wood (Pistia stratiotes) *ITS Eng. J* **6** 2301–9271
[13] Patra M and Sharma A 2000 Mercury toxicity in plants *Bot. Rev.* **66**
[14] Pescod M B 1973 Investigation of Rational Effluent and Stream Standard for Tropical Countries. *Enviromental Eng. Div. Asian Inst. Technol. Bangkok. Bangkok.*
[15] Barakwan R A, Trihadiningrum Y and Bagastyo A Y 2019 Characterization of alum sludge from Surabaya Water Treatment Plant, Indonesia *J. Ecol. Eng.* **20**
[16] Harahap A, Barus T A, Mulya M B and Ilyas S 2018 Macrozoobenthos diversity as bioindicator of water quality in the Bilah river, Rantauprapat *Journal of Physics: Conference Series* vol 1116
[17] Juhri D A 2017 Effect of heavy metals (cadmium, chromium, and lead) on wet weight reduction of water spinach (Ipomoea aquatica forsk) as extension material for vegetable farmers *Lantern Educ. Res. Cent. LPPM UM Metro.* **2** 219–29

AUTHOR INDEX

Institute of Physics
Dirac House, Temple Back
Bristol BS1 6BE UK

ISSN: 1742-6588
ISBN 978-1-7138-5097-7

CURRAN ASSOCIATES INC.
proceedings.com